MRC/CSO SPHSU
University of Glasgow
200 Renfield Street
Glasgow G2 3QB

General Editors

M. S. BARTLETT, F.R.S., *and* D. R. COX, F.R.S.

THE ANALYSIS OF BINARY DATA

The Analysis of Binary Data

D. R. COX

Department of Mathematics
Imperial College, London

LONDON

CHAPMAN AND HALL

First published 1970
by Methuen & Co Ltd

Reprinted 1977
by Chapman and Hall Ltd
11 *New Fetter Lane, London EC4P 4EE*
Reprinted 1980

© 1970 *D. R. Cox*

ISBN 0 412 15340 8

Printed in Great Britain
by Spottiswoode Ballantyne Ltd
Colchester and London

Distributed in the U.S.A. by Halsted Press,
a Division of John Wiley & Sons, Inc., New York

Preface

This monograph concerns the analysis of binary (or quantal) data, i.e. data in which an observation takes one of two possible forms, e.g. success or failure. The central problem is to study how the probability of success depends on explanatory variables and groupings of the material.

Many particular methods, especially significance tests, have been proposed for such problems and one of the the main themes of the monograph is that these methods are unified by considering models in which the logistic transform of the probability of success is a linear combination of unknown parameters. These linear logistic models play here much the same role as do normal-theory linear models in the analysis of continuously distributed data.

Some knowledge of the theory of statistics is assumed. I have written primarily for statisticians, but I hope also that scientists and technologists interested in applying statistical methods will, by concentrating on the examples, find something useful here.

I am very grateful to Dr Agnes M. Herzberg and Dr P. A. W. Lewis for extremely helpful comments. I acknowledge also the help of Mrs Jane Gentleman who programmed some of the calculations.

D. R. COX

London
April 1969

v

Contents

CHAPTER 1

Introduction

1.1 Binary data

Suppose that on each individual we have an observation that takes one of two possible forms. The following are examples:

  an electronic component may be defective, or may be non-defective;

  a test animal may die from a specified dose of a poison, or may survive;

  a subject may give the correct reply in an experimental situation, or may give a wrong reply;

  a test specimen may fracture when struck with a standardized blow, or may not;

and so on. If for the ith individual we can represent this observation, or response, by a random variable, Y_i, we may without loss of generality code the two possible values of Y_i by 1 and 0 and write

$$E(Y_i) = \text{prob}\,(Y_i = 1) = \theta_i, \text{prob}\,(Y_i = 0) = 1 - \theta_i, \quad (1.1)$$

say. It is often convenient to call $Y_i = 1$ a 'success' and $Y_i = 0$ a 'failure'. It is reasonable to call such observations *binary*; an older term is *quantal*.

We assume that such binary observations are available on n individuals, usually assumed to be independent. The problem is to develop good methods of analysis for assessing any dependence of θ_i on explanatory variables representing, for example, groupings of the individuals or quantitative regressor variables.

We have followed the usual terminology and have distinguished between (a) response variables and (b) explanatory or regressor variables, the variables of the second type being used to explain or predict variation in variables of the first type.

Sometimes a binary response variable arises by condensing a more complex response. Thus a component may be classed as defective when a quantitative test observation falls outside specification limits or, more generally, when a set of test observations falls in an unacceptable region. When this is done we need to consider whether there is likely to be serious loss in treating the problem in terms of a binary response.

In addition to a binary response variable there may be further response variables for each individual. Thus in a psychological experiment there may be available as well as the rightness or wrongness of a reply, the time taken to make the reply. Joint analysis of the response variables is then likely to be informative. We shall, however, in the remainder of the monograph concentrate on purely binary responses.

1.2 Some examples

It is convenient to begin with a few simple specific examples illustrating the problems to be considered.

Example 1.1 *The* 2×2 *contingency table.* Suppose that there are two groups of individuals, 1 and 2, of sizes n_1 and n_2 and that on each individual a binary response is obtained. The groups may, for example, correspond to two treatments. Suppose further that we provisionally base the analysis on the assumption that all individuals respond independently with probability of success depending only on the group, and equal, say, to $\theta^{(1)}$ and $\theta^{(2)}$ in the two groups.

Then the data can be reduced to the numbers of successes R_1 and R_2 in the two groups; in fact (R_1, R_2) form sufficient statistics for the unknown parameters $(\theta^{(1)}, \theta^{(2)})$. It is conventional to set out the numbers of successes and failures in a 2×2 contingency table as in Table 1.1. Nearly always it is helpful to calculate the proportions of successes in the two groups.

In Table 1.1 the rows refer to a dichotomy of the response variable, i.e. to a random variable in the mathematical model; the columns refer to a factor classifying the individuals into two groups, the classification being considered as non-random for the purpose of the analysis. It is possible to have contingency tables in which both rows and columns correspond to random variables.

As a specific example, the data of Table 1.2 refer to a retrospective survey of physicians (Cornfield, 1956). Data were obtained on a group of lung cancer patients and a comparable control group. The numbers of individuals in the two groups are approximately equal and are in no way representative of population frequencies. Hence it is reasonable to make an analysis conditionally on the observed total numbers in the two groups. An essential initial step is to calculate the sample proportions of successes in the two groups R_1/n_1 and R_2/n_2, and these are shown at the bottom of Table 1.2. Further analysis of the table is concerned with the precision of these proportions.

TABLE 1.1

A 2 × 2 contingency table

	Group 1	Group 2	Total
Successes	R_1	R_2	$R_1 + R_2$
Failures	$n_1 - R_1$	$n_2 - R_2$	$n_1 + n_2 - R_1 - R_2$
Total	n_1	n_2	$n_1 + n_2$
Proportion of successes	R_1/n_1	R_2/n_2	

TABLE 1.2

Numbers of smokers in two groups of physicians

	Controls	Lung cancer patients	
Smokers	32	60	92
Non-smokers	11	3	14
Total	43	63	106
Proportion of smokers	0·744	0·952	

When two groups are to be compared using binary observations, it will often be sensible to make an initial analysis from a 2 × 2 contingency table. However, the assumptions required to justify condensation of the data into such a form are not to be taken lightly. Thus in §5.2 we shall deal with the methods to be followed when pairs of individuals in the two groups are correlated. The most frequent inadequacy of an analysis by a single 2 × 2 contingency table is, however, the presence of further factors influencing the response, i.e. non-constancy of the probability of success within groups. To ignore such further factors can be very misleading; see Exercise 3 of Appendix A.

For many purposes, it is not necessary to give the numbers of failures in addition to the numbers of successes and the numbers of trials per group. In most of the more complex examples we shall therefore omit the second row of the 2 × 2 table.

There are many extensions of the 2 × 2 table; there follow some examples.

Example 1.2 *Several* 2 × 2 *contingency tables.* Suppose that to compare two treatments we have several sets of observations, each of the form of Example 1.1. The different sets may correspond to levels of a further factor or, as in the following specific example, may correspond to different blocks of an experimental design. In general, in the jth set of data, let n_{j1} and n_{j2} be the sample sizes in the two groups and let R_{j1} and R_{j2} be the total numbers of successes.

Table 1.3 (Gordon and Foss, 1966) illustrates this. On each of 18 days babies not crying at a specified time in a hospital ward served as subjects. On each day one baby chosen at random formed the experimental group and the remainder were controls. The binary response was whether the baby was crying or not at the end of a specified period. In Table 1.3, not-crying is taken as a 'success' and the observed numbers r_{j1} and r_{j2} are therefore the numbers of babies in the two groups not crying; the common convention is followed of denoting observed values of random variables by lower case letters. The special feature of this example is that $n_{j2} = 1$, so that R_{j2} takes values 0 and 1; usually there would be several individuals in each group.

The object of the analysis is to assess the effect of the treatment on the probability of success. The tentative basis for the analysis is that there is in some sense a constant treatment effect throughout the experiment, even though there may be some systematic variation from day to day. The experiment has the form of a randomized block design, in fact a matched pair design, but the binary nature of the response and the varying numbers of individuals in the groups complicate the analysis.

For the reasons indicated after Example 1.1 it would not in general be a sound method of analysis to pool the data over days, thus forming a single 2 × 2 contingency table with entries $\sum R_{j1}$, $\sum R_{j2}$, etc.

One simple, if approximate, method of analysis that is not distorted by systematic differences between groups is to calculate for the jth group the difference in the proportions of successes, i.e.

$$\frac{R_{j2}}{n_{j2}} - \frac{R_{j1}}{n_{j1}}. \tag{1.2}$$

TABLE 1.3

The crying of babies

Day	No. of control babies n_{j1}	No. not crying r_{j1}	No. of experimental babies n_{j2}	No. not crying r_{j2}
1	8	3	1	1
2	6	2	1	1
3	5	1	1	1
4	6	1	1	0
5	5	4	1	1
6	9	4	1	1
7	8	5	1	1
8	8	4	1	1
9	5	3	1	1
10	9	8	1	0
11	6	5	1	1
12	9	8	1	1
13	8	5	1	1
14	5	4	1	1
15	6	4	1	1
16	8	7	1	1
17	6	4	1	0
18	8	5	1	1

This is an unbiased estimate for the jth set of the difference between the probabilities of success. When (1.2) is averaged over all the sets, an unbiased estimate of the mean difference between groups results. A difficulty of this analysis is that the quantities (1.2) have in general different precisions for different j.

Later, further examples will be given whose analysis requires the combination of data from several 2×2 contingency tables. *Example* 1.3 *A* 2×2^m *system.* Suppose that there are m two-level factors thought to affect the probability of success. Let a binary response be observed on each individual and suppose that there is at least one individual corresponding to each of the 2^m cells, i.e. possible factor combinations; usually there would be an appreciable number of individuals in each cell.

It would be possible to think of such data as arranged in an $(m + 1)$-dimensional table, with two levels in each dimension. Alternatively and more usefully, we can think of a 2×2^m table in which the two rows correspond to success and failure and the 2^m columns are the 2^m standard treatments of the factorial system. One of the problems of analysis is to examine what order of interactions, in some suitable sense, is needed to represent the data; that is, we think first of 2^m probabilities of success, one for each cell, and then try to represent these usefully in terms of a smaller number of parameters.

Table 1.4 is a specific example of a 2×2^4 system based on an observational study by Lombard and Doering (1947) (see Dyke and Patterson (1952) for a detailed discussion of the analysis of this data). In this study, the response concerned individuals' knowledge of cancer, as measured in a test, a 'good' score being a success and a 'bad' score a failure. There were four factors expected to account for variation in the probability of success, the individuals being classified into 2^4 cells depending on presence or absence of exposure to A, newspapers; B, radio; C, solid reading; D, lectures. In Table 1.4 the standard notation for factor combinations in factorial experiments is used; thus ac denotes the cell in which A and C are at their upper levels and B and D at their lower levels.

Some general conclusions can be drawn from inspection of the cell proportions of successes. This is an example where both the response variable and the factors are reduced to two levels from a more complex form.

Example 1.4 *Serial order.* Suppose that a series of independent binary responses is observed and that it is suspected that the probability of success changes systematically with serial order. One application is to some types of data in experimental psychology, where a subject makes a series of responses, each either correct or incorrect, and where the probability of correct response is suspected of changing systematically. In this context, the treating of successive responses as independent may, however, be seriously misleading. Another application is in human genetics where each child in a family is classified as having or not having a particular genetic defect. Here, except for possible complications from multiple births, each family leads to a sequence of binary responses. In this application, data from many

TABLE 1.4

2×2^4 system. Study of cancer knowledge

	1	a	b	ab
Successes	84	75	13	35
Number of trials	477	231	63	94
Proportion of successes	0·176	0·325	0·206	0·372
	c	ac	bc	abc
Successes	67	201	16	102
Number of trials	150	378	32	169
Proportion of successes	0·447	0·532	0·500	0·604
	d	ad	bd	abd
Successes	2	7	4	8
Number of trials	12	13	7	12
Proportion of successes	0·167	0·538	0·571	0·667
	cd	acd	bcd	abcd
Successes	3	27	1	23
Number of trials	11	45	4	31
Proportion of successes	0·273	0·600	0·250	0·742

families would be required. We would then have a number of sequences, usually not all of the same length, and it would be required to examine the data for systematic changes, with serial order, in the probability of a genetic defect. It would not usually be reasonable to suppose that in the absence of such effects the probability of a defect is the same for all families.

Example 1.5 *Stimulus binary-response relation.* The following situation is of wide occurrence. There is a stimulus under the experimenter's control; each individual is assigned a level of the

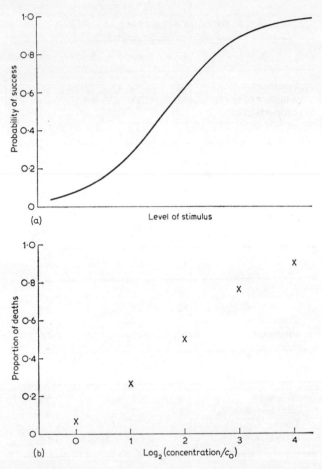

FIGURE 1.1 Stimulus–response curves. (*a*) Idealized theoretical curve. (*b*) Empirical curve from Table 1.5.

stimulus and a binary response then observed. One important field of application is bioassay, where, for example, different levels of stimulus may represent different doses of a poison, and the binary response is death or survival. Similar situations arise in many other fields.

In such applications it is often possible to choose a measure x of stimulus level such that the probability of success is zero for large negative x, unity for large positive x and is a strictly increasing function of x. In fact it has the mathematical properties of a continuous cumulative distribution function; see Fig. 1.1a. If the x scale is suitably chosen, the distribution function will be symmetric; for example, in the particular

TABLE 1.5

Simple form of bioassay

	Concentration				
	c_0	$2c_0$	$4c_0$	$8c_0$	$16c_0$
	Log_2 (Concentration/c_0)				
	0	1	2	3	4
No. of deaths	2	8	15	23	27
No. of individuals	30	30	30	30	30
Proportion of deaths	0·067	0·267	0·500	0·767	0·900

application mentioned above it is often helpful to take x as log dose.

Table 1.5 gives some illustrative data; at each of a number of dose levels a group of individuals is tested and the number dying recorded. In Fig. 1.1b the proportions dying are plotted against log concentration and, except for random fluctuations, give a curve similar to the idealized theoretical one of Fig. 1.1a. The object of the analysis of such data is to summarize and assess the properties of the stimulus–response curve. Sometimes, as in many bioassay problems, the aspect of primary importance is the level of x at which the probability of success is $\frac{1}{2}$, or sometimes the level x_p at which some other specified value, p, of the probability

2

of success is achieved. In other applications, for example in experimental psychology, the steepness of the response curve is the aspect of primary interest. We shall refer to x variously as a stimulus, or as a regressor or explanatory variable.

Example 1.6 *A* 2 × *g* *contingency table with trend.* Suppose that there are g groups of individuals, each individual having a binary response. If individuals' responses are independent, and the probability of success is constant within each group, we can, just as in the discussion of the 2 × 2 table, condense the data, giving simply the number of trials, n_j and the number of successes R_j for the jth group $(j = 1, ..., g)$. In contingency table form we then have Table 1.6.

TABLE 1.6

A 2 × *g* *contingency table*

	Group 1 ... Group g
Successes	R_1 ... R_g
Failures	$n_1 - R_1$ ... $n_g - R_g$
Total	n_1 ... n_g
Proportion of successes	R_1/n_1 ... R_g/n_g

Now suppose that the g groups are meaningfully ordered and that it is reasonable to expect that any change in the probability of success is monotonic with group order. If, further, scores $x_1, ..., x_g$ can be allocated to the g groups such that a smooth relation between the probability of success and the value of x is reasonable, the situation is then formally the same as that of Example 1.5. A distinction is that in the previous example the main interest is in the location and shape of a response curve, whereas in the present example there is usually substantial interest in the null hypothesis that the probability of success is constant.

Table 1.7 gives a specific example quoted by Armitage (1955). The data refer to children aged 0 to 15 and the binary response concerns whether a child is or is not a carrier for *Streptococcus pyogenes*; the children are grouped into three sets depending on tonsil size. No objective x variable was available and the tentative scores -1, 0, 1 were therefore assigned to the three groups. A difficulty in the interpretation of such data is that other possibly relevant explanatory variables, age, sex, etc. should be considered.

Example 1.7 *Multiple regression with binary response.* The previous three examples have involved a dependence between a probability of success and a single regressor or explanatory variable, x; they are thus comparable with normal-theory regression problems with a single regressor variable. Suppose

TABLE 1.7

Nasal carrier rate and tonsil size

	Tonsils present, not enlarged	Tonsils enlarged	
		+	++
Carriers	19	29	24
Total	516	589	293
Proportion of carriers	0·0368	0·0492	0·0819
x	-1	0	1

now that there are for each individual m explanatory variables $x_1, \ldots, x_m$, regarded as non-random, and a binary response. It is necessary to assess the relation between the probability of success and the variables $x_1, \ldots, x_m$.

This type of problem occurs especially in medical contexts. The binary response may represent success or failure of a particular treatment; death or survival over a specified time following treatment; or death from a particular cause, as contrasted with death from some other cause. The variables $x_1, \ldots, x_m$ represent quantitative, or qualitative, properties of an individual thought to influence the response; possible x variables are age, time since diagnosis, initial severity of symptoms, sex (scored as a zero-one variable), aspects of medical history of the individual, etc.

One particular example is a study of the factors affecting the probability of having coronary heart disease (see Walker and Duncan (1967) who give a number of earlier references). Another example concerns perinatal mortality (Feldstein, 1966). Data from an industrial problem involving two regressor variables are given in Section 5.5.

In some situations, especially with observational data, we need to analyse changes in the probability of success, there being a considerable number of potential explanatory or regressor variables. Two broad approaches are possible, illustrated by Examples 1.3 and 1.7.

In the first approach, the values of the explanatory variables are coarsely grouped, in the extreme case each explanatory variable taking only two values, as in Example 1.3. With m explanatory variables there will thus be at least 2^m cells, for each of which the proportion of successes can be found. An advantage of this approach is that quite complicated 'interactions' can be detected. Possible disadvantages stem from the necessity of coarse grouping and from the fact that if m is at all large many of the cells will either be empty or contain very few observations.

In the second approach, a regression-like model is taken, expressing a smooth and simple dependence of the probability of success on the values of the explanatory variables. No grouping is necessary, but a disadvantage is that relatively complicated interactions may be difficult to detect. In practice both methods are useful, separately and in combination. With both, problems concerning alternative choices of explanatory variables, familiar from normal-theory regression methods, arise in essentially the same form.

Example 1.8 *A binary time series.* In the previous examples it is a reasonable provisional assumption to suppose that the responses of different individuals are independent. In the analysis of a binary time series we are directly concerned with the lack of independence of different observations.

A specific example concerns daily rainfall. It is often reasonable first to classify days as wet (success) or dry (failure); there results a sequence of 1's and 0's, a binary time series. The amounts of rainfall on wet days can be analysed separately.

More generally, if we consider a response 1 as the occurrence of an event and a response 0 as non-occurrence, a binary time series is a series of events in discrete time. If, further, the proportion of 1's is low, the series approximates to a series of point events in continuous time; in particular, a completely random binary series, in which all responses have independently the same probability, θ, of giving a success, tends, as $\theta \to 0$, to a Poisson process. Cox and Lewis (1966) have summarized statistical techniques for the analysis of point events in continuous time.

In this section a number of relatively simple problems have been described which can be generalized in various ways:

(i) we may have situations of more complex structure. Thus, instead of the single response curve of Example 1.5, we might have several response curves and be interested in comparing their shapes;

(ii) we may have multivariate binary responses and consider problems analogous to those of multivariate normal theory;

(iii) we may have responses taking not just two possible values, but some small number greater than two. Can the techniques to be developed for analysing binary responses be extended?

Some of the more complex problems will be considered later. Others are indicated as exercises or further results in Appendix A.

As in other fields of statistical analysis, problems of two broad types arise. We require techniques for efficient analysis and assessment of uncertainty in the context of an assumed probabilistic model. Also we need techniques for tabular and graphical display and condensation of data, sometimes with the objective of finding a suitable model for more detailed analysis. This aspect is relatively more important with extensive data. This book deals mainly but by no means entirely with the first type of technique.

CHAPTER 2

The linear logistic model

2.1 Normal-theory linear models

For each of the problems with a binary response set out in Chapter 1 there is an analogous problem with a continuously distributed response. Particularly if the frequency distribution of the random component of the response is approximately normal, it would be sensible to think about using a regression or analysis of variance technique. Such techniques are based on the method of least squares, applied to an appropriate linear model. One of the objects of the present book is to set out a general approach to the analysis of binary data that is broadly comparable with the use of linear models with normally distributed responses.

It is therefore convenient to review briefly the relevant portions of normal-theory. Consider then the analysis of data represented by random variables $Y_1, ..., Y_n$, now not in general binary, it being supposed provisionally that

$$E(Y_i) = \sum_{s=1}^{p} a_{is}\beta_s, \tag{2.1}$$

where $\{a_{is}\}$ $(i = 1, ..., n; s = 1, ..., p)$ are known constants and $\beta_1, ..., \beta_p$ are unknown parameters. In an obvious matrix notation

$$E(\mathbf{Y}) = \mathbf{a}\boldsymbol{\beta}, \quad E(Y_i) = \mathbf{a}_i\boldsymbol{\beta} \qquad (i = 1, ..., n), \tag{2.2}$$

where $\mathbf{a}_i$ is the ith row of the matrix $\mathbf{a}$; the random variables and unknown parameters are given respectively by the column vectors $\mathbf{Y}$ and $\boldsymbol{\beta}$.

The least squares estimates $\hat{\boldsymbol{\beta}}$ of $\boldsymbol{\beta}$ satisfy the simultaneous linear equations

$$(\mathbf{a}'\mathbf{a})\hat{\boldsymbol{\beta}} = \mathbf{T}, \tag{2.3}$$

where

$$\mathbf{T} = \mathbf{a}'\mathbf{Y}, \qquad T_s = \sum_{i=1}^{n} a_{is} Y_i, \qquad (2.4)$$

and $\mathbf{a}'$ denotes the transpose of $\mathbf{a}$. The estimates $\hat{\boldsymbol{\beta}}$ can be regarded as minimizing, with respect to $\boldsymbol{\beta}^*$, the sum of squares

$$(\mathbf{Y} - \mathbf{a}\boldsymbol{\beta}^*)'(\mathbf{Y} - \mathbf{a}\boldsymbol{\beta}^*).$$

We consider also vectors of fitted values $\hat{\mathbf{Y}} = \mathbf{a}\hat{\boldsymbol{\beta}}$, and of residuals

$$\mathbf{Y} - \hat{\mathbf{Y}}, \qquad (2.5)$$

and the residual sum of squares

$$S_{\text{res}} = (\mathbf{Y} - \hat{\mathbf{Y}})'(\mathbf{Y} - \hat{\mathbf{Y}}). \qquad (2.6)$$

Part of the attraction of an analysis based on $\hat{\boldsymbol{\beta}}$, S_{res} and on the residuals, arises from the descriptive reasonableness of fitting a model by least squares. From a much more theoretical point of view, very strong optimum properties follow if it is assumed that

$Y_1, \ldots, Y_n$ are independently normally distributed with constant variance, say σ^2. $\qquad (2.7)$

For then $(\mathbf{T}, S_{\text{res}})$ are sufficient statistics for the unknown parameters $(\boldsymbol{\beta}, \sigma^2)$. From the sufficiency it follows that:

(i) so long as the analysis is based on (2.1) and (2.7), all the information about the unknown parameters is contained in $(\mathbf{T}, S_{\text{res}})$ and no other function of the Y_i's needs to be considered;

(ii) the adequacy of (2.1) and (2.7) is examined in principle from the conditional distribution of $\mathbf{Y}$ given $(\mathbf{T}, S_{\text{res}})$ which is, under the model, free of the parameters $(\boldsymbol{\beta}, \sigma^2)$. This can be shown to be equivalent to using, implicitly or explicitly, some aspects of the configuration of the residuals.

Suppose now that we replace (2.7) by the much weaker assumption that

$$\text{cov}(Y_i, Y_j) = 0, \qquad \text{var}(Y_i) = \sigma^2 \qquad (2.8)$$
$$(i \neq j = 1, \ldots, n).$$

The least squares estimates, using the Y_i's only through the quantities $\mathbf{T}$, then have minimum variance among all unbiased estimates that are linear functions of the Y_i's. We call the theory based on (2.8) the second-order theory. The books by Scheffé (1959), Plackett (1960) and Rao (1965) contain extensive accounts of the above results and their developments.

The linear model (2.1) allows the inclusion under this theory of the statistical techniques of analysis of variance and covariance, of simple and multiple regression analysis, and of the fitting of polynomials in one or more variables and of linear combinations of harmonic functions of known wavelength.

2.2 Application of least squares to binary data

We now return to binary data and consider methods for assessing a dependence of $\theta_i = \text{prob}(Y_i = 1)$ on explanatory variables or groupings of the data.

One simple possibility is to retain (2.1) and to consider the model

$$\theta_i = \text{prob}(Y_i = 1) = E(Y_i) = \sum_{s=1}^{p} a_{is}\beta_s, \qquad (2.9)$$

and to apply the method of least squares directly to the binary observations, i.e. to treat the observations 0 and 1 just as if they were quantitative observations. This is a computationally simple method which can be useful, particularly when only relatively small changes in θ_i are involved.

The method has the following limitations. Since Y_i takes only the values 0 and 1, $Y_i^2 = Y_i$ and

$$\text{var}(Y_i) = \theta_i(1 - \theta_i). \qquad (2.10)$$

That is, the condition (2.8) of constant variance, required even for the second-order theory of least squares, cannot hold, except in the rather uninteresting case when the θ_i's are all the same. However, it is known that quite appreciable changes in $\text{var}(Y_i)$ induce only a modest loss of efficiency. Further, at least in the range, say, $0 \cdot 2 \leqslant \theta_i \leqslant 0 \cdot 8$, the function $\theta_i(1 - \theta_i)$ changes relatively little. Therefore, within this range, there is unlikely to be a serious loss of efficiency arising from the changes in $\text{var}(Y_i)$.

The average variance of the Y_i's can be estimated approximately from the data in various ways, for example by

$$\text{ave}(Y_i)\{1 - \text{ave}(Y_i)\}, \qquad (2.11)$$

where $\text{ave}(Y_i)$ is the overall proportion of 1's in the whole data.

The estimate (2.11) can thus be used as an approximate theoretical residual mean square in the standard analysis of variance based on (2.9); a comparison with it of an empirical residual mean square obtained from (2.6) provides a test of the adequacy of the model (2.9).

However, if the values of $\theta_i(1 - \theta_i)$ vary appreciably with i, there may be a serious loss of information in using the unweighted least squares estimates (2.3). One possible development is to use an iterative scheme in which fitted values $\hat{Y}_i$ are obtained from (2.9) and then weights $\{\hat{Y}_i(1 - \hat{Y}_i)\}^{-1}$ are used in a weighted least squares analysis. This is directly related to a maximum likelihood analysis of the model (2.9); see Exercise 2.

A second closely related matter is that because the Y_i's are not normally distributed, no method of estimation that is linear in the Y_i's will in general be fully efficient.

The most serious restriction on the usefulness of (2.9) arises, however, from the condition

$$0 \leqslant \theta_i \leqslant 1. \tag{2.12}$$

It is possible for the least squares estimates obtained from (2.3) to lead to a vector of fitted values $\mathbf{a}\hat{\boldsymbol{\beta}}$, some components of which do not satisfy (2.12). In such cases it would be reasonable to consider modified least squares estimates obtained by minimizing with respect to $\boldsymbol{\beta}^*$ the sum of squares

$$(\mathbf{Y} - \mathbf{a}\boldsymbol{\beta}^*)'(\mathbf{Y} - \mathbf{a}\boldsymbol{\beta}^*),$$

subject to the constraints that all elements of the vector $\mathbf{a}\boldsymbol{\beta}^*$ satisfy (2.12). The resulting mathematical programming problem is computationally considerably more complex than the unmodified least squares calculation, whenever the unmodified solution fails to satisfy (2.12).

Quite apart from the computational difficulties induced by (2.12), there is the more serious matter that the parameters in the model (2.9) inevitably have a limited interpretation and range of validity. Consider for instance the analysis of the stimulus–response curve of Example 1.5 and suppose that we take a linear model (2.9) in which the probability of success is a linear function of the variable x. Even if the data were satisfactorily linear over the observed range of stimuli, it is certain that the linear relation will fail outside a restricted range of stimuli; further it is usually rather unlikely that the underlying relationship has an 'inclined plane' form, i.e. a range with zero probability of success, a linear portion of positive slope and a range with unit probability of success; the general shape illustrated in Fig. 1.1a is much more common. The use of a linear model is likely to mean that even small-scale extrapolation is hazardous and that a different experimenter, using a different range of stimuli, would be likely to find an apparently different stimulus–response relationship. Similar remarks apply to the other examples of §1.2.

Of course, all postulated models have at best limited and approximate validity. The point here is that the use of a model, the nature of whose limitations can be foreseen, is not wise, except for very limited purposes.

We now turn to a type of model in which the constraint (2.12) is automatically satisfied.

2.3 The linear logistic model

In many respects the simplest way of representing the dependence of a probability on explanatory variables so that the constraint (2.12) is inevitably satisfied, is to postulate a dependence for $i = 1, \ldots, n$,

$$\theta_i = \frac{e^{a_i\beta}}{1 + e^{a_i\beta}}, \tag{2.13}$$

$$1 - \theta_i = \frac{1}{1 + e^{a_i\beta}}. \tag{2.14}$$

Here, as in (2.2), a_i is a row of known constants and β is a column of unknown parameters. Equations (2.13) and (2.14) are equivalent to

$$\lambda_i = \log\left(\frac{\theta_i}{1 - \theta_i}\right) = a_i\,\beta = \sum_{s=1}^{p} a_{is}\beta_s, \tag{2.15}$$

or, collecting the n values $(i = 1, \ldots, n)$ together, we can write

$$\lambda = a\beta. \tag{2.16}$$

We call $\lambda_i = \log\{\theta_i/(1 - \theta_i)\}$, the logistic transform of the probability θ_i, and (2.16) a linear logistic model. Equation (2.16) is to be compared with (2.2). An alternative name for λ_i is log odds.

In many ways (2.16) is the most useful analogue for binary response data of the linear model (2.2) for normally distributed data. From a rather formal point of view, one analogy arises from considerations of sufficiency.

To see this, let $Y_1, \ldots, Y_n$ be independent binary random variables distributed in accordance with (2.16) and let $y_1, \ldots, y_n$ denote the corresponding observed values. Then the likelihood contains a factor (2.13) whenever $y_i = 1$ and a factor (2.14) whenever $y_i = 0$. Thus the likelihood is

$$\frac{\prod_{i=1}^{n} e^{\mathbf{a}_i \boldsymbol{\beta} y_i}}{\prod_{i=1}^{n} (1 + e^{\mathbf{a}_i \boldsymbol{\beta}})} = \frac{\exp\left(\sum_{s=1}^{p} \beta_s t_s\right)}{\prod_{i=1}^{n} (1 + e^{\mathbf{a}_i \boldsymbol{\beta}})}, \qquad (2.17)$$

where $T_s = \sum a_{is} Y_i$ is as given by (2.4) and t_s is its observed value. Since the Y_i's are binary, T_s is a random subtotal of the sth column of the matrix $\mathbf{a}$, the elements included in the sum being those that correspond to successes, $Y_i = 1$.

Thus there are simple sufficient statistics T_s ($s = 1, \ldots, p$) which are moreover formally identical, except for the absence of the residual sum of squares, with those for the normal-theory linear model. One consequence is that methods of analysis based on the linear logistic model are often more closely related to normal-theory techniques than might have been anticipated.

2.4 The 2×2 contingency table

We now return to some of the examples given in §1.2 and indicate possibly appropriate linear logistic models. Throughout, θ_i denotes a probability of success and λ_i the corresponding logistic transform, $\log\{\theta_i/(1 - \theta_i)\}$; in general developments we use $\boldsymbol{\beta}$ for the vector of unknown parameters in the linear logistic model, but in particular cases other Greek symbols may be used.

Consider first the 2×2 contingency table, Example 1.1, in which the basis of the analysis is that there are just two different values for the probability of success, namely $\theta^{(1)}$ and $\theta^{(2)}$, corresponding to the two groups of observations. The corresponding linear logistic model is

$$\lambda^{(1)} = \alpha, \qquad \lambda^{(2)} = \alpha + \varDelta. \qquad (2.18)$$

Here Δ represents the difference between groups on a logistic scale. Note that

$$\Delta = \lambda^{(2)} - \lambda^{(1)}$$

$$= \log\left[\frac{\theta^{(2)}\{1 - \theta^{(1)}\}}{\{1 - \theta^{(2)}\}\theta^{(1)}}\right]. \qquad (2.19)$$

Thus e^{Δ} is the ratio of the odds of success versus failure in the two groups.

Now (2.18) is a direct re-parametrization of the problem from $(\theta^{(1)}, \theta^{(2)})$, taking values in the unit square, to (α, Δ), taking values in the whole plane. That is, no additional empirical assumptions about the data are involved beyond those already made in the representation with the pair of binomial parameters $(\theta^{(1)}, \theta^{(2)})$. It is convenient to say that the model (2.18) is saturated with parameters; that is, the number of independent parameters equals the number of independent binomial probabilities. The parametrization (2.18) is fruitful if Δ is in some sense a good way of characterizing the difference between the two groups.

Now there are many ways in which that difference might be measured. Thus if $h(\theta)$ is a monotonic increasing function of θ, we could take $h\{\theta^{(2)}\} - h\{\theta^{(1)}\}$, or some function of it, as such a measure. Possible choices of $h(\theta)$ are the following:

(i) $h(\theta) = \theta$, leading to $\theta^{(2)} - \theta^{(1)}$;
(ii) $h(\theta) = \log\theta$, leading to $\log\{\theta^{(2)}/\theta^{(1)}\}$ or to $\theta^{(2)}/\theta^{(1)}$;
(iii) $h(\theta) = \log(1 - \theta)$, leading to $\log[\{1 - \theta^{(1)}\}/\{1 - \theta^{(2)}\}]$ or
      to $\{1 - \theta^{(1)}\}/\{1 - \theta^{(2)}\}$;
(iv) $h(\theta) = \log\{\theta/(1 - \theta)\}$, leading to Δ of (2.19).

There are several criteria for choosing between alternative ways of assessing the difference between groups.

(a) Quite often it will be reasonable to require that if successes and failure are interchanged the measure of the difference between groups is either unaltered or at most changed in sign.

(b) It is desirable, wherever possible, to work with a measure that will remain stable over a range of conditions.

(c) If a particular measure of difference has an especially clear-cut direct practical meaning, for example an explicit economic interpretation, we may decide to use it.

(d) A measure of the difference for which the statistical theory of inference is simple is, other things being equal, a good thing.

Now consideration (a) excludes the simple ratio measures (ii) and (iii). Note, however, that when θ is very small, (iv) is equivalent to (ii), whereas when θ is very near 1, (iv) is equivalent to (iii).

Requirement (b) is particularly important. Ideally we would have the comparison of the two groups made under a range of circumstances, i.e. there would be a set of 2×2 tables in which the overall proportion of successes changed appreciably from set to set. It would then be an empirical matter to find that measure of difference, e.g. that function $h(\theta)$, if any, which is stable. By finding a scale on which the difference between groups is stable, there is economy in summarization and in addition the conclusions can be applied more confidently to a fresh set of conditions. In the absence of empirical data to establish the appropriate $h(\theta)$, it is worth noting that for the logistic model with a fixed Δ we can vary α in (2.18) arbitrarily and produce a family of situations in which the overall proportion of successes is arbitrary. On the other hand, in the corresponding representation directly in terms of probabilities, a given difference $\delta = \theta^{(2)} - \theta^{(1)}$ is consistent with only a limited range of values for $\theta^{(2)}$ and $\theta^{(1)}$ individually. This makes it plausible, but not inevitable, that the parameter Δ in the logistic model (2.18) for the analysis of a single 2×2 table has a broader range of validity than the parameter $\delta = \theta^{(2)} - \theta^{(1)}$.

Requirement (c) may in some situations justify an analysis in terms of $\theta^{(2)} - \theta^{(1)}$. If, say, the two groups represent alternative industrial processes and $\theta^{(1)}$, $\theta^{(2)}$ are the probabilities of defectives, then $m\{\theta^{(2)} - \theta^{(1)}\}$ will be approximately the difference between the numbers of defectives produced in a large number m of trials, and this may have an economic interpretation. Finally, while the requirement of simplicity of inference should not be of over-riding importance, we shall find that 'small-sample' theory of inference about the logistic parameter Δ is particularly simple.

The above discussion deals with the comparison of success rates of two groups, success or failure being a response regarded as varying randomly between individuals and the divisions into two groups being non-random. There is, however, a different situation also involving a comparison of success rates and for which a comparison on a logistic scale is particularly appropriate.

Suppose that we have a population in which for each individual there are two binary random variables U and W. There are thus four types of individual corresponding to $U = W = 0$; $U = 1$, $W = 0$; $U = 0$, $W = 1$; $U = W = 1$, the corresponding probabilities being respectively π_{00}, π_{10}, π_{01}, π_{11}. For example, we might have

$$U = \begin{cases} 1 \text{ for non-smokers,} \\ 0 \text{ for smokers,} \end{cases} \quad W = \begin{cases} 1 \text{ for sufferers from lung cancer,} \\ 0 \text{ for non-sufferers from lung} \\ \quad \text{cancer.} \end{cases}$$

Suppose now that we are interested in comparing the probabilities that $W = 1$ for two groups of individuals, corresponding respectively to $U = 0$ and to $U = 1$. There are at least three sampling procedures by which data may be obtained to study this.

First a random sample may be obtained from the whole population, allowing estimation of any function of the π_{ij}'s. In particular, the conditional probabilities that $W = 1$, given respectively $U = 0$ and $U = 1$, are

$$\frac{\pi_{01}}{\pi_{00} + \pi_{01}} \quad \text{and} \quad \frac{\pi_{11}}{\pi_{10} + \pi_{11}}, \qquad (2.20)$$

and the difference of the logistic transforms of these is

$$\log \left(\frac{\pi_{11}}{\pi_{10}} \right) - \log \left(\frac{\pi_{01}}{\pi_{00}} \right) = \log \left(\frac{\pi_{11} \pi_{00}}{\pi_{10} \pi_{01}} \right). \qquad (2.21)$$

Secondly, it may be possible to draw two random samples, one from each of the sub-populations $U = 0$ and $U = 1$, the sizes of the samples being fixed and having no connection with the marginal probabilities $\pi_{00} + \pi_{01}$ and $\pi_{10} + \pi_{11}$. Thus, in the example, equal-sized samples of smokers and non-smokers might be taken and the values of W obtained in a prospective study. Then the individual π_{ij}'s cannot be estimated, but only functions of the conditional probabilities of W given $U = 0$ and $U = 1$, i.e. the probabilities (2.20). In particular the logistic difference (2.21) can be estimated.

Thirdly, the roles of U and W may be interchanged and random samples taken from the sub-populations $W = 0$ and $W = 1$. Thus, in the example, a retrospective study might take equal-sized samples of lung cancer patients and of controls and find the value of U for each patient. In such a study the quantities that can be estimated are functions of

$$\frac{\pi_{10}}{\pi_{00} + \pi_{10}} \quad \text{and} \quad \frac{\pi_{11}}{\pi_{01} + \pi_{11}}. \quad (2.22)$$

Now the logistic differences of (2.22) and (2.20) are the same, namely (2.21), and hence the logistic difference can be estimated formally in the same way from either prospective or retrospective studies. Suppose, however, that we wanted the difference

$$\text{prob}\,(W = 1 \,|\, U = 0) - \text{prob}\,(W = 1 \,|\, U = 1) =$$

$$\frac{\pi_{01}}{\pi_{00} + \pi_{01}} - \frac{\pi_{11}}{\pi_{10} + \pi_{11}};$$

this can be estimated from (2.20) but not, without further major assumptions, from (2.22).

Thus, to summarize, if we are ultimately interested in regarding W as a response and U as a conditioning variable, but in fact the sampling is based on the inverse scheme of fixing W and observing U, the difference on a logistic scale, but not on any other, can be estimated.

2.5 Some more complex examples

Consider now Example 1.2, where there are k independent 2×2 tables comparing two groups 1 and 2 under a range of conditions. Possible linear logistic models are that if $\theta_j^{(1)}$ and $\theta_j^{(2)}$ denote the probabilities of success in the two groups for the jth table, then, for $j = 1, \ldots, k$,

$$\lambda_j^{(1)} = \alpha_j, \qquad \lambda_j^{(2)} = \alpha_j + \Delta_j, \qquad (2.23)$$

or

$$\lambda_j^{(1)} = \alpha_j, \qquad \lambda_j^{(2)} = \alpha_j + \Delta, \qquad (2.24)$$

or

$$\lambda_j^{(1)} = \alpha, \qquad \lambda_j^{(2)} = \alpha + \Delta. \qquad (2.25)$$

Equation (2.23) is a re-parametrization of the model in which the probabilities $\theta_j^{(1)}$ and $\theta_j^{(2)}$ are arbitrary; under (2.24) there are arbitrary differences between the sets of data, but a constant logistic difference between groups; under (2.25) the sets are homogeneous. In many situations, including the special example of Table 1.3, the most useful basis for an analysis is (2.24). We are likely to be interested in inference about the parameter Δ, allowing the nuisance parameters $\alpha_1, \ldots, \alpha_k$ to take care of differences between sets; in particular the null hypothesis $\Delta = 0$ will often be of special concern. The model (2.24) is the direct analogue of a randomized block model in normal theory, the parameter Δ representing a treatment effect and the parameters $\alpha_1, \ldots, \alpha_k$ block effects.

Consider next the 2×2^m system of Example 1.3. For simplicity we write down formulae for $m = 2$. There are four probabilities $\theta^{(00)}$, $\theta^{(01)}$, $\theta^{(10)}$, $\theta^{(11)}$ corresponding to the four treatments in the 2×2 factorial system. If we write

$$\lambda^{(00)} = \mu - \tfrac{1}{2}\tau_1 - \tfrac{1}{2}\tau_2 + \tfrac{1}{2}\tau_{12}, \qquad \lambda^{(01)} = \mu - \tfrac{1}{2}\tau_1 + \tfrac{1}{2}\tau_2 - \tfrac{1}{2}\tau_{12},$$
$$\lambda^{(10)} = \mu + \tfrac{1}{2}\tau_1 - \tfrac{1}{2}\tau_2 - \tfrac{1}{2}\tau_{12}, \qquad \lambda^{(11)} = \mu + \tfrac{1}{2}\tau_1 + \tfrac{1}{2}\tau_2 + \tfrac{1}{2}\tau_{12},$$

$$(2.26)$$

we have a direct re-parametrization in which no new empirical assumptions are involved. This model is saturated with parameters, because there are four parameters for four binomial probabilities. The parameters τ_1, τ_2 and τ_{12} measure on a logistic scale the main effects of the two factors and their interaction; μ is the overall mean. The parameters $(\mu, \tau_1, \tau_2, \tau_{12})$ can take any values in a four-dimensional parameter space. Note, however, that if we define contrasts directly as linear combinations of the θ's rather than of the λ's, the resulting parameters are constrained by the requirement that all the θ's lie between 0 and 1. The arguments for defining the contrasts on the logistic scale parallel those of §2.4.

In general, with the 2×2^m system, we can set out a direct re-parametrization in terms of a general mean and $2^m - 1$ factorial contrasts.

Now a representation in terms of factorial contrasts is usually fruitful only if some or all of the parameters representing higher-order interactions are negligible. In the simple case, $m = 2$, we will be interested in the model without an interaction term:

$$\lambda^{(00)} = \mu - \tfrac{1}{2}\tau_1 - \tfrac{1}{2}\tau_2, \qquad \lambda^{(01)} = \mu - \tfrac{1}{2}\tau_1 + \tfrac{1}{2}\tau_2,$$
$$\lambda^{(10)} = \mu + \tfrac{1}{2}\tau_1 - \tfrac{1}{2}\tau_2, \qquad \lambda^{(11)} = \mu + \tfrac{1}{2}\tau_1 + \tfrac{1}{2}\tau_2. \qquad (2.27)$$

Thus we shall be interested in testing the hypothesis $\tau_{12} = 0$ in (2.26), in inference about τ_1 and τ_2 in (2.27) and possibly in testing, say, $\tau_1 = 0$ in (2.27).

The formulation of models and of the types of question to be asked is, for the logistic model, exactly parallel to that for the normal-theory linear model.

Examples 1.4 to 1.6 all involve the dependence of the probability of success on a single regressor variable, x. In the simplest logistic model

$$\lambda_i = \alpha + \beta x_i, \qquad (2.28)$$

$$\theta_i = \frac{e^{\alpha + \beta x_i}}{1 + e^{\alpha + \beta x_i}}. \qquad (2.29)$$

If θ_i given by (2.29) is plotted as a function of x_i, a symmetrical stimulus–response curve of the type shown in Fig. 1.1a is obtained, provided that $\beta > 0$. If $\beta < 0$ the curve is monotonic decreasing; the case $\beta = 0$ serves as a null hypothesis in which the probability of success is constant. The numerical value of β measures the steepness of the response curve. The value of x at which the probability of success is $\frac{1}{2}$ is $-\alpha/\beta$. In Example 1.4, the serial order problem, x_i is the serial number of the ith observation, i.e. in a single series $x_i = i$. A reasonable basis for an analysis in which there are several series from the same source would often be to assign a separate parameter α to each series, but to assume provisionally that the parameter β is common to all series. This corresponds to the fitting of parallel lines to sets of data in normal-theory linear regression.

Similar remarks apply to the more complex multiple regression problem, Example 1.7.

To sum up, the linear logistic expression (2.16) provides a flexible collection of models that are at least potentially capable of representing a range of situations involving binary responses. Broadly, the linear logistic models, and the questions they are intended to answer, are analogous to the normal-theory linear models, which underlie the techniques of analysis of variance and covariance, and simple and multiple regression. Of course, not all problems about approximately normally distributed observations are best tackled via a linear model; correspondingly it is wrong to think that a linear logistic model will be successful every time binary responses are encountered.

2.6 Some numerical properties of the logistic curve

In order to appreciate the practical meaning of an analysis on the logistic scale, it is important to appreciate the meaning of

3

differences on a logistic scale. To help in this, the following
properties are useful.

The first column of Table 2.1 gives the standardized logistic
function

$$\Lambda(x) = e^x/(1 + e^x). \tag{2.30}$$

This is helpful in interpreting the parameters of the response
curve (2.29). Since $\Lambda(1) = 0\cdot731$, the parameter β in the form

$$\frac{e^{\alpha+\beta x}}{1 + e^{\alpha+\beta x}}$$

is such that $1/\beta$ is approximately the distance in x units between
the 75% point and the 50% point of the curve. Similarly the
distance between the 95% point and the 50% point is approximately $3/\beta$, etc.

The difference $\Delta\theta$ of two probabilities corresponding to a
difference $\Delta\lambda$ on a logistic scale depends on the particular values
of the probabilities involved. Approximately, however, we have
on differentiating (2.15) that

$$\Delta\lambda \sim \frac{\Delta\theta}{\theta(1-\theta)}. \tag{2.31}$$

Thus near $\theta = \frac{1}{2}$, $\Delta\lambda \sim 4\Delta\theta$, whereas for small θ, $\Delta\lambda \sim \Delta\theta/\theta$,
and for θ very near 1, $\Delta\lambda \sim \Delta\theta/(1-\theta)$. Alternatively, differences
on a logistic scale can be interpreted easily by thinking from the
beginning not in terms of the probability θ of success but rather
in terms of the odds for success against failure, namely $\theta/(1-\theta)$.
The exponential of a logistic difference is thus a ratio of odds.
If $\Delta\lambda = 1$, the odds in the second situation are about $2\cdot7$ times
those in the first situation.

2.7 Alternative scales

In §2.2 we considered briefly a linear model formulated directly
in terms of probabilities θ_i, whereas the remainder of the chapter
has concentrated on models linear on a logistic scale. It is, however, possible to consider models linear on other scales and here
we illustrate two such, briefly, in terms of a stimulus–response
curve.

Since the object here is only to compare the mathematical form of different transformations we can, without loss of generality, suppose that the 50% point is at $x = 0$ and we therefore write the logistic curve as

$$\frac{e^{\beta x}}{1 + e^{\beta x}}. \qquad (2.32)$$

For comparison, suppose that the probability of success varies linearly with x according to the function

$$\begin{cases} 1 & (x > \tfrac{1}{2}\gamma^{-1}), \\ \tfrac{1}{2} + \gamma x & (|x| \leqslant \tfrac{1}{2}\gamma^{-1}), \\ 0 & (x < -\tfrac{1}{2}\gamma^{-1}). \end{cases} \qquad (2.33)$$

Two further relationships correspond to the integrated normal and the angular transforms. According to the first, the probability of success at stimulus x is

$$\Phi(\xi x), \qquad (2.34)$$

where $\Phi(t)$ is the standardized normal integral,

$$\Phi(t) = \frac{1}{\sqrt{(2\pi)}} \int_{-\infty}^{t} e^{-\frac{1}{2}u^2} du.$$

When the angular transform is used, we take the probability of success at stimulus x to be

$$\begin{cases} 1 & (x > \tfrac{1}{4}\pi/\eta), \\ \sin^2(\eta x + \tfrac{1}{4}\pi) & (|x| \leqslant \tfrac{1}{4}\pi/\eta), \\ 0 & (x < -\tfrac{1}{4}\pi/\eta). \end{cases} \qquad (2.35)$$

In (2.32) to (2.35), the adjustable parameters β, γ, ξ and η allow the representation of relationships of different slopes.

If $\theta(x)$ denotes the probability of success at level x, the linearizing transformations associated with the response curves (2.32) to (2.35) are respectively from $\theta(x)$ to

$$\log\left\{\frac{\theta(x)}{1 - \theta(x)}\right\}, \quad \theta(x), \quad \Phi^{-1}\{\theta(x)\}, \quad \sin^{-1}\{\sqrt{\theta(x)}\}. \quad (2.36)$$

To compare the four response curves, all of which are symmetric, we have to choose the scale constants β, γ, ξ and η in comparable form. This has been done rather arbitrarily by taking $\beta = 1$ and then choosing γ, ξ and η so that all four curves agree at the 80% point. We get $\gamma = 0.144$, $\xi = 0.607$ and $\eta = 0.232$. Table 2.1 gives the resulting functions; they all behave symmetrically about $x = 0$.

For most purposes the logistic and the normal agree closely over the whole range. The only exception is when special interest attaches to the regions where the probability of success is either very small, or very near one. Then the normal curve approaches its limit more rapidly than the logistic. The angular and linear relations agree with the other two reasonably well

TABLE 2.1

Comparison of probability of success as given by four stimulus–response curves

x	Logistic	Normal	Angular	Linear
0	0·500	0·500	0·500	0·500
0·5	0·622	0·619	0·615	0·608
1·0	0·731	0·728	0·724	0·716
1·5	0·818	0·818	0·821	0·825
2·0	0·881	0·887	0·900	0·933
2·5	0·924	0·935	0·958	1·000
3·0	0·953	0·965	0·992	1·000
3·5	0·971	0·983	1·000	1·000
4·0	0·982	0·992	1·000	1·000
4·5	0·989	0·997	1·000	1·000
5·0	0·993	0·999	1·000	1·000

when the probability of success is in the range 0·1 to 0·9. Outside this range the finite limits on the last two curves usually seriously restrict their usefulness. Subject to these provisos, an analysis in terms of any of the four relations is likely to give virtually equivalent results (Claringbold, Biggers and Emmens, 1953; Naylor, 1964).

We have considered the general use of a model linear in the probabilities in §2.2. The statistical usefulness of the angular transformation arises from its variance-stabilizing property; see Exercise 8. With a number of groups of observations of equal size, fitting by unweighted least squares is appropriate. Both linear and angular functions are, however, severely limited for general usefulness by their finite range. The use of the integrated normal curve in connection with binary data has been extensively discussed (Finney, 1952). The absence of sufficient statistics is a theoretical disadvantage, primarily with relatively simple problems, where a simple solution is often available for the logistic and not for the integrated normal.

The above discussion is in terms of a stimulus–response curve. The general formulation would be in terms of linear models in which one or other of the transforms (2.36) is specified by a linear expression in unknown parameters. Thus for the 2×2 contingency table, the model based on the integrated normal analogous to (2.18) for the logistic is

$$\Phi^{-1}\{\theta^{(1)}\} = \alpha', \qquad \Phi^{-1}\{\theta^{(2)}\} = \alpha' + \Delta'. \qquad (2.37)$$

The close numerical equivalence between the integrated normal and the logistic functions assures us that, unless the probabilities are very near 0 or 1, (2.37) and (2.18) are virtually equivalent, with

$$\alpha' \simeq \frac{\alpha}{0\cdot607}, \qquad \Delta' \simeq \frac{\Delta}{0\cdot607}.$$

Nevertheless, the theoretical analysis of (2.37) is more complicated and less elegant than that of the logistic analogue (2.18).

CHAPTER 3

The empirical logistic transform

3.1 Introduction

We now consider an important special situation in which the method of weighted least squares can be used to give asymptotically efficient procedures for the linear logistic model. We suppose that the binary observations are grouped into sets, all trials in the same set having the same probability of success under the model. Each set is assumed to contain a reasonably large number of trials. The 2×2 contingency table, Example 1.1, is a rather extreme case with only two sets. The analysis of stimulus–response data, Example 1.5, is also of this type, provided that the stimulus variable x takes a limited number of values with an appreciable number of replicate observations at each value.

Example 1.4, the serial order problem, is a case where the methods of the present chapter are not directly applicable, because each trial has a different probability of success under the simplest linear logistic model. Also a multiple regression analysis of observational data with an appreciable number of independent variables is unlikely to yield data that can be grouped into sets with many observations in each set.

Suppose then that in the jth set of observations the probability of success is constant and equal say to $\theta^{(j)}$ ($j = 1, \ldots, g$). Let there be n_j trials in the set, R_j of these successes. Then so long as we base the analysis on the assumption that the probability of success is constant within a set, we can disregard the individual responses and work only with the R_j, which are binomially distributed with

$$E(R_j/n_j) = \theta^{(j)}, \qquad \text{var}(R_j/n_j) = \theta^{(j)}\{1 - \theta^{(j)}\}/n_j. \quad (3.1)$$

Now suppose that the linear logistic model specifies that

$$\lambda^{(j)} = \log[\theta^{(j)}/\{1 - \theta^{(j)}\}]$$

is a linear combination of a $p \times 1$ vector of parameters, $\boldsymbol{\beta}$. We write

$$\boldsymbol{\lambda} = \mathbf{c}\boldsymbol{\beta}, \qquad \lambda^{(J)} = \sum_{s=1}^{p} c_{js}\beta_s, \qquad (3.2)$$

where $\mathbf{c}$ is a $g \times p$ matrix. Equation (3.2) is a condensed form of (2.16). In (2.16), however, $\boldsymbol{\lambda}$ is an $n \times 1$ vector, where n is the number of binary responses, whereas in (3.2) $\boldsymbol{\lambda}$ is a $g \times 1$ vector, where g is the number of groups. Likewise the matrix $\mathbf{c}$ is a condensed form of $\mathbf{a}$, with one row for each group of binary observations.

The method that follows applies equally to linear relations based on quite general transforms of the probabilities. We can obtain some extra generality by considering a model

$$\boldsymbol{\psi}(\boldsymbol{\theta}) = \mathbf{c}\boldsymbol{\beta}, \qquad (3.3)$$

where $\boldsymbol{\psi}(\boldsymbol{\theta})$ denotes the vector whose components are $\psi\{\theta^{(J)}\}$. For $\psi(x) = \log\{x/(1-x)\}$ this becomes (3.2); other special cases are discussed in §2.7. We proceed to 'linearize' the observed proportions by a corresponding transformation.

Now for large n_j, provided that $\theta^{(J)}$ is not too near 0 or 1,

$$Z'_j = \psi(R_j/n_j) \qquad (3.4)$$

is nearly normally distributed. As $n_j \to \infty$ the asymptotic mean and variance are respectively

$$\psi\{\theta^{(J)}\} \quad \text{and} \quad [\psi'\{\theta^{(J)}\}]^2 \, \theta^{(J)}\{1-\theta^{(J)}\}/n_j. \qquad (3.5)$$

Moreover, the asymptotic variance in (3.5) is consistently estimated by

$$V'_j = \left\{\psi'\left(\frac{R_j}{n_j}\right)\right\}^2 \frac{R_j(n_j - R_j)}{n_j^3}, \qquad (3.6)$$

i.e. by replacing $\theta^{(J)}$ by R_j/n_j.

The method of least squares can now be applied to $Z'_1, \ldots, Z'_g$ using empirically estimated weights from (3.6). Alternatively, if the V'_j's do not vary much, unweighted least squares can be used as an approximation.

The weighted analysis is equivalent to treating $Z'_j/\sqrt{V'_j}$ as having unit variance with

$$E\left(\frac{Z'_j}{\sqrt{V'_j}}\right) = \sum_{s=1}^{p} \frac{c_{js}}{\sqrt{V'_j}}\beta_s, \qquad (3.7)$$

where we ignore random errors in $\sqrt{V'_j}$ on the right-hand side. We set up least squares equations, tests, etc in the usual way; the fact that the theoretical variance is unity allows goodness of fit to be tested by comparing the empirical residual mean square with its theoretical value. The whole procedure is justified asymptotically as all $n_j \to \infty$, because then the Z'_j's are asymptotically normally distributed, the random variation in the V'_j's being a second-order effect; the strong theoretical optimum properties of least squares methods applied to normally distributed observations (§2.1) thus hold in the limit. Extensive trials for the special problem of the stimulus–response curve (Example 1.5) suggest that the method gives good estimates even when the n_j's are quite small (Berkson, 1953); there is, however, need for further work to delimit more explicitly when the method can be safely used, especially for significance tests and confidence limits.

The approach can be described as the method of least squares with empirically estimated weights as applied to suitably transformed observations. For the particular case of the logistic stimulus–response curve, Berkson (1953) introduced the method and called it minimum logit chi-squared, because of an analogy between the sum of squares to be minimized and a chi-squared goodness of fit statistic.

When $\psi(x) = \sin^{-1}\sqrt{x}$, the angular transformation discussed briefly in §2.7, we have from (3.4) and (3.6) that

$$Z'_j = \sin^{-1}\sqrt{\left(\frac{R_j}{n_j}\right)} \quad \text{and} \quad V'_j = \frac{1}{4n_j}. \tag{3.8}$$

In particular, when the n_j are all equal, or nearly so, unweighted least squares is appropriate, applied directly to the Z'_j. The resulting simplicity is the primary justification for considering the angular transformation. As noted in §2.7 the finite range of the transformation is a limitation on its general usefullness.

For the logistic transformation, (3.4) and (3.6) become

$$Z'_j = \log\left(\frac{R_j}{n_j - R_j}\right), \qquad V'_j = \frac{n_j}{R_j(n_j - R_j)}. \tag{3.9}$$

These we now consider in more detail; we call Z'_j in (3.9) the empirical logistic transform of (R_j, n_j).

3.2 The modified transform

We now concentrate on the logistic transform. The version (3.9) needs modification if only because it is undefined at $R_j = 0$ or n_j. In extensive data occasional extreme values of R_j are to be expected, even if on the whole the conditions for large-sample theory apply.

One plausible way of modifying (3.9) (Haldane, 1955; Anscombe, 1956) is to define a transform as

$$Z_j(a) = \log\left(\frac{R_j + a}{n_j - R_j + a}\right) \tag{3.10}$$

and then to choose the constant a so that the expected value of (3.10) is as nearly as possible $\lambda^{(j)} = \log[\theta^{(j)}/\{1 - \theta^{(j)}\}]$. It is convenient for the remainder of the section to omit the suffix and superfix j.

Write $R = n\theta + U\sqrt{n}$, where

$$E(U) = 0, \qquad E(U^2) = \theta(1 - \theta)$$

and U is of order one in probability as $n \to \infty$. Then

$$Z(a) - \lambda = \log\left\{1 + \frac{U}{\theta\sqrt{n}} + \frac{a}{\theta n}\right\} - \log\left\{1 - \frac{U}{(1 - \theta)\sqrt{n}} + \frac{a}{(1 - \theta)n}\right\}$$

$$= \frac{U}{\theta(1 - \theta)\sqrt{n}} + \frac{a(1 - 2\theta)}{\theta(1 - \theta)n} - \frac{(1 - 2\theta)U^2}{2\theta^2(1 - \theta)^2 n} + o\left(\frac{1}{n}\right), \tag{3.11}$$

where the terms neglected in (3.11) are of smaller order than $1/n$ in probability. Thus

$$E\{Z(a)\} - \lambda = \frac{(1 - 2\theta)(a - \frac{1}{2})}{\theta(1 - \theta)n} + o\left(\frac{1}{n}\right) \tag{3.12}$$

and therefore an appropriate choice is $a = \frac{1}{2}$. It is interesting, but not essential for the argument, that we can nullify the term in $1/n$ in (3.12) by a single choice of a, independent of θ.

We thus put

$$Z = \log\left(\frac{R + \frac{1}{2}}{n - R + \frac{1}{2}}\right). \tag{3.13}$$

A number of corresponding modifications of the large-sample variance V' have been suggested. Gart and Zweifel (1967) have investigated their properties by methods analogous to those used above and their work suggests the definition

$$V = \frac{(n+1)(n+2)}{n(R+1)(n-R+1)}. \quad (3.14)$$

This is such that to a close approximation

$$\operatorname{var}(Z) = E(V), \quad (3.15)$$

so that V is a nearly unbiased estimate of $\operatorname{var}(Z)$.

TABLE 3.1

Exact expectation of the modified logistic transform, Z

			$E(Z)$	
θ	λ	$n = 5$	10	20
0·5	0·000	0·000	0·000	0·000
0·6	0·405	0·401	0·407	0·406
0·7	0·847	0·821	0·850	0·849
0·8	1·386	1·280	1·375	1·389
0·9	2·197	1·798	2·062	2·179
0·92	2·442	1·910	2·229	2·398
0·94	2·751	2·026	2·409	2·651
0·96	3·178	2·146	2·604	2·946
0·98	3·892	2·270	2·815	3·295
0·99	4·595	2·334	2·927	3·494

The result $E(Z) = \lambda$ is still an approximate large-sample one even though the order of approximation is closer than that for the 'crude' transform (3.9). Table 3.1 gives the exact expectation of Z as a function of θ and n. Provided that $n(1 - \theta) > 1$, and, for small θ, $n\theta > 1$, the agreement between $E(Z)$ and λ is good enough for most purposes.

The usefulness of these results is broadly as follows. The linear combination $\sum l_j Z_j$, where the l_j's are constants, is such that to a close approximation

$$E(\sum l_j Z_j) = \sum l_j \lambda^{(j)}, \qquad \operatorname{var}(\sum l_j Z_j) = E(\sum l_j^2 V_j). \quad (3.16)$$

Note that the condition of approximate unbiasedness of V_j

means that, at least when many sets are combined, the variance estimate, $\sum l_j^2 V_j$, in (3.16) will be a good one.

Finally, note that the calculation leading to (3.12) and the choice $a = \frac{1}{2}$ are relevant only if the Z_j's are to be used in combinations with constant weights and therefore does not apply when the least squares procedure with empirical weights, derived from (3.7), is used; for then linear combinations of the Z_j's themselves do not arise. We return to that case in §3.5 and in the meantime discuss situations where the theory of the present section is relevant.

3.3 The 2 × 2 contingency table

The simplest situation to which the theory of §3.2 applies is the 2×2 contingency table, Example 1.1. Here we have just two binomial probabilities, corresponding to the two sets. We define for $j = 1, 2$

$$Z_j = \log\left(\frac{R_j + \frac{1}{2}}{n_j - R_j + \frac{1}{2}}\right), \tag{3.17}$$

$$V_j = \frac{(n_j + 1)(n_j + 2)}{n_j(R_j + 1)(n_j - R_j + 1)}. \tag{3.18}$$

Hence, under the saturated model (2.18), $Z_2 - Z_1$ is an estimate of the logistic difference Δ and has a standard error approximately $\sqrt{(V_1 + V_2)}$. Hence, using the approximate normality of the distribution of $Z_2 - Z_1$, the null hypothesis $\Delta = 0$ can be tested, and confidence intervals for Δ calculated.

Because the logistic model is saturated, i.e. has as many logistic parameters as there are binomial probabilities, the unweighted combination $Z_2 - Z_1$ is the unique estimate of Δ from the least squares analysis. Thus the distinction between unweighted and weighted least squares does not arise.

Example 3.1 *Comparison of two groups of physicians.* We now apply the results to the data of Table 1.2, for which $n_1 = 43$ and $n_2 = 63$, and for which the observed values of R_1 and R_2 are $r_1 = 32$ and $r_2 = 60$. Thus, with z and v denoting the observed values of Z and V, we have that

$$z_1 = \log(32\cdot5/11\cdot5) = 1\cdot039, \ v_1 = 0\cdot116;$$

$$z_2 = \log(60\cdot5/3\cdot5) = 2\cdot850, \ v_2 = 0\cdot271.$$

We treat $z_2 - z_1 = 1 \cdot 811$ as an observed value of a random variable having mean $\varDelta$ and standard deviation $\sqrt{(0 \cdot 116 + 0 \cdot 271)} = 0 \cdot 622$. Therefore the two-sided significance level in testing $\varDelta = 0$ corresponds to a standardized normal deviate of $2 \cdot 91$ and so to $P \simeq 0 \cdot 004$.

Approximate 95% confidence limits for $\varDelta$ are $(0 \cdot 592, 3 \cdot 03)$. Later we shall compare these values with those determined by other methods.

Consider briefly the extension to the analysis of several 2×2 tables, Example 1.2. The method just outlined can be used on the k individual 2×2 tables to give estimates $\tilde{\varDelta}_1, ..., \tilde{\varDelta}_k$ of the logistic differences in the individual tables, each with its approximate standard error. Note, however, that we could hardly apply large-sample methods to the numerical data of Table 1.3 because of the very small sample sizes.

Using the estimates, we can examine a number of questions about the parameters $\varDelta_1, ..., \varDelta_k$. We defer a discussion of these, however, to §5.3, where some 'exact' procedures are given. Note that in any case the use of weighted combinations of $\tilde{\varDelta}_1, ..., \tilde{\varDelta}_k$ would normally be appropriate, so that, as noted at the end of §3.2, the modified forms (3.13) and (3.14) of the logistic transform would not be relevant.

3.4 An application to factorial experiments

A more complex illustration of the use of the modified transform (3.13) and of (3.14) arises with the 2×2^m factorial system of Example 1.3 and more generally with factorial systems with more than two levels.

In the 2×2^m system, provided that there is an appreciable number of observations in each cell, we can consider 2^m modified logistic transforms, Z_j and associated variance estimates V_j, one pair for each cell. Suppose now either that:

(i) we analyze the data in terms of a saturated logistic model in which main effects and interactions of all orders, 2^m parameters in all, are included;

or

(ii) we use a non-saturated logistic model, for example one containing only main effects, but make for simplicity an unweighted least squares analysis.

In both cases we are led to compute estimates of factorial contrasts by the standard formulae. Further, the total for each contrast will be the sum of 2^m terms, each term a Z_j with coefficient ± 1. Hence all estimated contrasts will have the same variance which will be estimated by $\sum V_j$.

Note that in (3.16) the contrasts estimated have equal precision only because an unweighted analysis is used. Further, the requirement that V_j should be a nearly unbiased estimate of $\mathrm{var}(Z_j)$ is particularly appropriate when the final variance of interest is, as here, $\sum \mathrm{var}(Z_j)$.

If factors at more than two levels are involved, meaningful contrasts corresponding to single degrees of freedom are, if possible, isolated and standardized to have unit variance.

There are a number of techniques associated with the normal-theory analysis of factorial experiments that can now be applied. These include the half-normal plotting technique (Daniel, 1959) in which the absolute values of the contrasts are ranked in increasing order and plotted against the expected values of the order statistics from a semi-normal population. The known variance of the contrasts enables a theoretical line to be drawn on the plot. The adequacy of the approximations involved in applying this to contrasts based on logistic transforms have been confirmed by sampling experiments (Cox and Lauh, 1967).

The use of these methods is best illustrated by a numerical example.

Example 3.2 *A* $2 \times (3 \times 2^2)$ *factorial experiment.* Table 3.2 gives the data from $2 \times (3 \times 2^2)$ experiment comparing two detergents, a new product X and a standard product M (Ries and Smith, 1963). The three factors are water softness, at three levels, temperature, at two levels, and a factor whose two levels correspond to previous experience and no previous experience with M. For each of the twelve factor combinations, a number, n_j, of individuals, between 48 and 116, use both detergents and r_j of these prefer X, the remainder preferring M.

To construct a set of single degrees of freedom, we suppose that the three levels of water softness are in some sense equally spaced and so split all contrasts involving softness into linear and quadratic components. The quantities z_j and v_j computed by (3.13) and (3.14), are given in Table 3.2. Next, standardized contrasts are computed. For example, for the quadratic main effect of softness, we compute first

$$(0 \cdot 477 + 0 \cdot 332 - 0 \cdot 337 - 0 \cdot 575) -$$
$$2(0 \cdot 275 + 0 \cdot 355 - 0 \cdot 156 - 0 \cdot 704) +$$
$$(0 \cdot 171 + 0 \cdot 070 + 0 \cdot 150 - 0 \cdot 414) = 0 \cdot 334. \qquad (3.19)$$

TABLE 3.2

Number r_j of preferences for brand X out of n_j individuals

		M previous non-user		M previous user	
		Temperature		Temperature	
Water softness		Low	High	Low	High
Hard	r_j	68	42	37	24
	n_j	110	72	89	67
	z_j	0·477	0·332	−0·337	−0·575
	v_j	0·0381	0·0563	0·0457	0·0637
Medium	r_j	66	33	47	23
	n_j	116	56	102	70
	z_j	0·275	0·355	−0·156	−0·704
	v_j	0·0348	0·0723	0·0391	0·0634
Soft	r_j	63	29	57	19
	n_j	116	56	106	48
	z_j	0·171	0·070	0·150	−0·414
	v_j	0·0344	0·0703	0·0376	0·0851

The estimated variance is

$$(0 \cdot 0381 + \cdots + 0 \cdot 0637) +$$
$$4(0 \cdot 0348 + \cdots + 0 \cdot 0634) +$$
$$(0 \cdot 0344 + \cdots + 0 \cdot 0851) = 1 \cdot 270,$$

and finally the standardized contrast is

$$\frac{0 \cdot 334}{\sqrt{1 \cdot 270}} = 0 \cdot 296;$$

this has approximately unit theoretical variance. The contrasts
are collected in Table 3.3.

A preliminary plot, or inspection of Table 3.3, shows that the
main effect of M is very highly significant; note that in Table 3.2,

z_j for a non-user cell is always greater than z_j for the corresponding user cell. Therefore the main effect of M was omitted from the plot and Fig. 3.1 shows the result; the line corresponds to the theoretical unit variance. There are three suspicious points namely, in order from the highest: T, $M \times S_L$ and $M \times T$. While these are of borderline significance individually, the fact that three of the large contrasts involve M and that this is a factor corresponding to a classification of the experimental units (individuals) rather than to a treatment, suggests splitting the

TABLE 3.3

Logistic factorial standardized
contrasts estimated from the
data of table 3.2

Temperature T	−1·894
User vs. non-user M	−4·642
Softness, linear S_L	−0·122
quadratic S_Q	0·296
$T \times M$	−1·479
$T \times S_L$	0·429
$T \times S_Q$	0·099
$M \times S_L$	−1·852
$M \times S_Q$	0·669
$T \times M \times S_L$	0·563
$T \times M \times S_Q$	0·621

experiment on factor M, i.e., analyzing separately the results from the two levels of M. The good agreement of the remaining points with the theoretical line is some check on the absence of additional components of error.

To do this, we analyze separately the two halves of Table 3.2 corresponding to the two levels of M. Note that a positive z_j corresponds to an average preference for X over M, and a negative z_j to an average preference for M over X. The conclusions are briefly as follows. For previous non-users of M, all cells show an average preference for X, i.e. positive z_j. There is no systematic change with temperature, but a decrease in preference with water softness, the average difference in z_j between soft and hard being $-0·284 \pm 0·223$, i.e. the approximate standard error estimated from the v_j's is 0·223. For previous users of M, all cells except one show an average preference for M, i.e. a negative z_j. The preference is stronger at the higher temperature, the average difference of z_j between high and low temperatures being $-0 \cdot 450 \pm 0 \cdot 193$. There is a general decrease in strength of preference for M with water softness; in terms of z_j this is an effect of opposite sign from that observed with previous non-users of M. The average difference in z_j between soft and hard

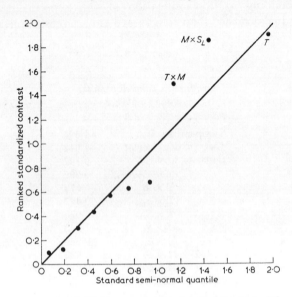

FIGURE 3.1 Half-normal plot for logistic factorial contrasts from Table 3.2, omitting main effect of M. Line gives theoretical slope.

for previous users of M is $0 \cdot 324 \pm 0 \cdot 241$. Thus in both cases the effects of softness, while suggestive, are not clearly established.

The above are broadly the conclusions reached by Ries and Smith (1963) by a series of chi-squared tests. The present approach, however, allows the estimation as well as the significance testing of effects.

The rather informal graphical approach is likely to be relatively more useful in more complex systems. Then some rough preliminary analysis will usually be necessary before fitting a suitably simplified model either by maximum likelihood (Dyke and Patterson, 1952) or by weighted least squares.

3.5 A weighted analysis

As remarked a number of times in the previous discussion, the modified transform (3.13) introduced in §3.2 is inappropriate when an empirically weighted analysis is used. For then the combinations of observations entering into the estimates are not linear functions of the Z_j's, so that the unbiasedness requirement for Z_j itself is irrelevant.

For special linear models it may be possible to develop especially appropriate transforms. However, one way of defining a modified transform that is at least plausible for the general model (3.7) is as follows. Each element of the matrix of coefficients on the left-hand side of the formal least squares equations derived from (3.7) consists of linear combinations of the $1/V_j'$'s; in fact the $(s,t)^{\text{th}}$ element is

$$\sum_{j=1}^{g} \frac{c_{js}c_{jt}}{V_j'}. \tag{3.20}$$

Next on the right-hand side of the formal least squares equations we have for the sth element

$$\sum_{j=1}^{g} c_{js} \frac{Z_j'}{V_j'}. \tag{3.21}$$

Now it seems reasonable, although no very precise formal justification will be given, to replace (3.20) and (3.21) by quantities whose expectations are very nearly the corresponding theoretical values

$$\sum_{j=1}^{g} c_{js} c_{jt} n_j \theta^{(j)}\{1 - \theta^{(j)}\} \tag{3.22}$$

and

$$\sum_{j=1}^{g} c_{js} n_j \theta^{(j)}\{1 - \theta^{(j)}\} \log\left\{\frac{\theta^{(j)}}{1 - \theta^{(j)}}\right\}. \tag{3.23}$$

The estimating equation will then be nearly an unbiased one (Durbin, 1960) and, at any rate if g is large, a consistent and efficient estimate will result. Now it is easily shown that for $n_j > 1$

$$E\left\{\frac{R_j(n_j - R_j)}{(n_j - 1)}\right\} = n_j \theta^{(j)}\{1 - \theta^{(j)}\} \tag{3.24}$$

4

and, by the methods used in §3.2, that approximately

$$E\left\{\frac{R_j(n_j - R_j)}{(n_j - 1)}\log\left(\frac{R_j - \frac{1}{2}}{n_j - R_j - \frac{1}{2}}\right)\right\}$$

$$= n_j\,\theta^{(J)}\{1 - \theta^{(J)}\}\log\left\{\frac{\theta^{(J)}}{1 - \theta^{(J)}}\right\}.$$

This leads us to define the random variables

$$B_{st} = \sum_{j=1}^{g} c_{js} c_{jt}\frac{R_j(n_j - R_j)}{(n_j - 1)}, \tag{3.25}$$

$$U_s = \sum_{j=1}^{g} c_{Js}\frac{R_j(n_j - R_j)}{(n_j - 1)}\log\left(\frac{R_j - \frac{1}{2}}{n_j - R_j - \frac{1}{2}}\right) \tag{3.26}$$

and to consider the estimating equations

$$\sum B_{st}\,\tilde{\theta}_t^{(w)} = U_s. \tag{3.27}$$

Note that the contribution to U_s is 0 if $R_j = 0$ or n_j. The co-variance matrix of the estimates is asymptotically $\mathbf{B}^{-1}$ and the residual sum of squares corresponds, under the model, to a theoretical variance of unity.

These results are valid asymptotically as all the n_j's become large. Extensive investigations by Berkson (1955a) of the stimulus–response curve, Example 1.5, with just two unknown parameters show that the method using the crude transform and weights Z'_j, V'_j gives good point estimates even when the group sizes are small. He did not examine the adequacy of the standard errors calculated from least squares theory. In §§6.2 and 6.3 these ideas are developed further.

CHAPTER 4

Exact analysis for a
single parameter

4.1 Introduction

In the previous chapter, we gave a method allowing the analysis of the general logistic model by the method of weighted least squares. There are two limitations to this. First, it can only be used when the observations are grouped into sets with a constant probability in each set, there being an appreciable number of observations in each set. Secondly, it is difficult to assess the adequacy of the approximations involved in the method and the magnitude of any loss of information arising from its efficiency being only asymptotic.

We now turn to methods derived from first principles directly from the linear logistic model; see Lehmann (1959, pp. 59, 134) for some corresponding results on the general exponential family of distributions. When we are interested in one of the logistic regression parameters, regarding the remainder as nuisance parameters, 'exact' fully efficient methods of analysis are in principle possible. That is, under the assumed logistic linear model, the methods have exactly known probabilistic properties and moreover are theoretically most sensitive in a certain sense. While too much weight should not be attached to these properties, which are concerned with what happens under an inevitably idealized model, nevertheless the methods of analysis are, at least in some important special cases, simple and informative.

Briefly, we shall be led to consider tests and confidence limits based on certain conditional distributions. Some general comments on the procedures are as follows:

(i) quite often, in relatively simple problems, the significance test of the null hypothesis that a regression coefficient is zero is easy both theoretically and computationally;

(*ii*) while the calculation of confidence limits is, in such cases, equally simple in theory, the computation tends to be more difficult;

(*iii*) in more complex problems, the formation of the appropriate conditional distribution may be difficult, or the distribution may even be degenerate;

(*iv*) the method is confined to the study of a single regression coefficient, regarding all the other parameters as nuisance parameters.

For the general development we shall use the notation of §2.3, in particular denoting the vector of unknown regression coefficients by $\boldsymbol{\beta}$. For individual problems, however, a different notation is often useful. Random variables are denoted by capital letters and, where it is essential to distinguish between them and the corresponding observed values, the latter are denoted by lower case letters. Often, however, it is not necessary to make the distinction explicitly.

4.2 Some general theory

(*i*) *Formulation.* We return to the linear logistic model in the general form set out in §2.3. The model for the independent binary random variables $Y_1, \ldots, Y_n$ is

$$\lambda_i = \log \left(\frac{\theta_i}{1 - \theta_i} \right) = \mathbf{a}_i \boldsymbol{\beta} = \sum_{s=1}^{p} a_{is} \beta_s. \qquad (4.1)$$

The likelihood of an observed binary sequence $y_1, \ldots, y_n$ is, as in (2.17),

$$\text{prob} (Y_1 = y_1, \ldots, Y_n = y_n) = \frac{\exp \left(\sum\limits_{s=1}^{p} \beta_s t_s \right)}{\prod\limits_{i=1}^{n} (1 + e^{\mathbf{a}_i \boldsymbol{\beta}})}, \qquad (4.2)$$

where

$$t_s = \sum_{i=1}^{n} a_{is} y_i \qquad (4.3)$$

is the observed value of the random variable $T_s = \sum a_{is} Y_s$. The sufficient statistics $t_1, \ldots, t_p$ are sub-totals of the columns of the $n \times p$ matrix $\mathbf{a}$ formed from the rows $\mathbf{a}_i$, the elements summed corresponding to the rows in which successes occur.

We shall need the distribution of the random variables

$T_1, \ldots, T_p$. This follows directly from (4.2) on summing over all binary sequences that generate the particular values $t_1, \ldots, t_p$. In fact

$$\text{prob}\,(T_1 = t_1, \ldots, T_p = t_p) = \frac{c(t_1, \ldots, t_p)\exp\left(\sum\limits_{s=1}^{p} \beta_s t_s\right)}{\prod\limits_{i=1}^{n}(1 + e^{a_i\beta})}, \qquad (4.4)$$

where $c(t_1, \ldots, t_p)$ is the number of distinct binary sequences that yield the specified values $t_1, \ldots, t_p$. A generating function for $c(t_1, \ldots, t_p)$ is

$$C(\zeta_1, \ldots, \zeta_p) = \sum c(t_1, \ldots, t_p)\, \zeta_1^{t_1} \ldots \zeta_p^{t_p}$$

$$= \prod\limits_{i=1}^{n}(1 + \zeta_1^{a_{i1}} \ldots \zeta_p^{a_{ip}}). \qquad (4.5)$$

(*ii*) *Conditional inference.* Now suppose that we are interested in one of the regression parameters, regarding the remainder as nuisance parameters. Without loss of generality we suppose the parameter of interest to be β_p, the nuisance parameters being $\beta_1, \ldots, \beta_{p-1}$. To summarize what the data tell us about β_p we consider the conditional distribution of T_p given the observed values of $T_1, \ldots, T_{p-1}$. There are two lines of argument that lead to this. In both approaches we need consider only methods involving the sufficient statistics. First, we may look for a distribution that depends on the value of β_p, but not on the nuisance parameters $\beta_1, \ldots, \beta_{p-1}$. For any fixed and known value of β_p, a sufficient statistic for the remaining parameters is $(T_1, \ldots, T_{p-1})$. Therefore the distribution of the observations, and hence also of T_p, given that $T_1 = t_1, \ldots, T_{p-1} = t_{p-1}$, cannot involve $\beta_1, \ldots, \beta_{p-1}$. Further it can be shown (Lehmann, 1959, p. 130) that the use of this conditional distribution is the only way to obtain exact independence of the nuisance parameters. This is the standard Neyman-Pearson approach to the elimination of nuisance parameters. It leaves open the possibility that a more sensitive analysis of the data may be available by adopting a less drastic requirement than that the probability properties should be completely independent of the nuisance parameters.

The second approach to the conditional distribution (Fisher, 1956, §IV.4) stems from an idea which is conceptually very appealing in special cases, but difficult to formalize in general. This is that if we were given only the values $t_1, \ldots, t_{p-1}$, no conclusions could be drawn about β_p. The values $(t_1, \ldots, t_{p-1})$ determine the precision with which conclusions about β_p can be drawn and it is thus appropriate to argue conditionally on the observed values. This is to ensure that we attach to the conclusions the precision actually achieved and not that to be achieved hypothetically in a recognizably distinct situation that has in fact not occurred.

Fortunately, the two lines of argument point to the same conclusion in the present situation.

(*iii*) *Properties of conditional distributions.* To find the conditional distribution of T_p given $T_1 = t_1, \ldots, T_{p-1} = t_{p-1}$, we have that

$$\text{prob}\,(T_p = t_p | T_1 = t_1, \ldots, T_{p-1} = t_{p-1}) =$$

$$\frac{\text{prob}\,(T_1 = t_1, \ldots, T_p = t_p)}{\text{prob}\,(T_1 = t_1, \ldots, T_{p-1} = t_{p-1})}. \tag{4.6}$$

The numerator is given by (4.4) and the denominator by summing (4.4) over all possible t_p. Hence, when we form the ratio (4.6), several factors cancel and the conditional probability is

$$\frac{c(t_1, \ldots, t_p)\, e^{\beta_p t_p}}{\sum_u c(t_1, \ldots, t_{p-1}, u)\, e^{\beta_p u}}. \tag{4.7}$$

Note that (4.7) does not involve $\beta_1, \ldots, \beta_{p-1}$; indeed this fact is a consequence of the sufficiency of the conditioning statistics.

Most of the subsequent discussion centres on (4.7). It is convenient to simplify the notation. We denote the parameter β_p of interest by β, the statistic t_p by t and the conditioning statistic $(t_1, \ldots, t_{p-1})$ by $\mathbf{t}_{p-1}$. The distribution (4.7) can then be written

$$p_T(t; \beta) = \frac{c(\mathbf{t}_{p-1}, t)\, e^{\beta t}}{\sum_u c(\mathbf{t}_{p-1}, u)\, e^{\beta u}}. \tag{4.8}$$

An important special case of (4.8) corresponds to $\beta = 0$,

$$p_T(t; 0) = \frac{c(\mathbf{t}_{p-1}, t)}{\sum_u c(\mathbf{t}_{p-1}, u)}, \tag{4.9}$$

so that the distribution is determined by the combinatorial coefficients.

We denote the moment generating function of (4.8) by

$$M_T(z;\beta) = \sum_t e^{zt} p_T(t;\beta)$$

$$= \frac{\sum_t c(t_{p-1}, t) e^{(\beta+z)t}}{\sum_u c(t_{p-1}, u) e^{\beta u}}. \qquad (4.10)$$

In particular, when $\beta = 0$,

$$M_T(z;0) = \frac{\sum_t c(t_{p-1}, t) e^{zt}}{\sum_u c(t_{p-1}, u)}. \qquad (4.11)$$

Thus

$$M_T(z;\beta) = \frac{M_T(z+\beta;0)}{M_T(\beta;0)}, \qquad (4.12)$$

or, in terms of cumulant generating functions,

$$K_T(z;\beta) = K_T(z+\beta;0) - K_T(\beta;0), \qquad (4.13)$$

where $K_T(z;\beta) = \log M_T(z;\beta)$. The potential usefulness of (4.12) and (4.13) lies in the possibility of obtaining properties of the distribution of T for non-zero β from properties for $\beta = 0$.

(iv) *Optimum properties.* To complete the formal theoretical analysis, consider the problem of testing the null hypothesis $\beta = \beta_0$ against an alternative $\beta = \beta'$, with, say, $\beta' > \beta_0$. We confine attention to the sufficient statistics and, for the reasons outlined in (ii), consider the conditional distribution (4.8). To obtain a most powerful test we therefore apply the Neyman-Pearson lemma to (4.8), i.e. form a critical region from those sample points having large values of the likelihood ratio

$$\frac{p_T(t;\beta')}{p_T(t;\beta_0)} \propto e^{(\beta'-\beta_0)t}, \qquad (4.14)$$

the factor of proportionality being independent of t. Thus, for all $\beta' > \beta_0$, the critical region should consist of the upper tail of values of t and the resulting procedure is uniformly most powerful.

Corresponding to an observed value t, the one-sided significance level P_+ against alternatives $\beta > \beta_0$ is thus

$$P_+(t;\beta_0) = \sum_{u \geqslant t} p_T(u;\beta_0) \qquad (4.15)$$

and against alternatives $\beta < \beta_0$ the level is, correspondingly,

$$P_-(t;\beta_0) = \sum_{u \leq t} p_T(u;\beta_0). \qquad (4.16)$$

For two-sided alternatives it is usual to quote the value

$$P(t;\beta_0) = 2 \min \{P_+(t;\beta_0), P_-(t;\beta_0)\}. \qquad (4.17)$$

To obtain an upper $(1 - \epsilon)$ confidence limit for β, corresponding to an observed value t, we take the largest value of β, say $\beta_+(t)$, which is just significantly too large at the ϵ level, i.e. the solution of

$$\epsilon = P_-\{t;\beta_+(t)\}. \qquad (4.18)$$

Similarly a lower confidence limit is obtained from

$$\epsilon = P_+\{t;\beta_-(t)\}. \qquad (4.19)$$

Thus, for a given ϵ, only a discrete set of values is possible for the confidence limits, because the possible values of t are discrete. This can be shown to imply that if the true value of β equals one of the attainable limits, then the probability of error is the nominal value ϵ, and that otherwise the probability of error is less than ϵ. The confidence coefficient for the interval $(\beta_-(t), \beta_+(t))$ is thus at least $(1 - 2\epsilon)$. Note that if we are interested in the consistency of the data with a particular value of β, say β_0, we are concerned with (4.15) to (4.17), rather than with trying to achieve an arbitrary pre-assigned value of ϵ.

4.3 The 2×2 contingency table

The simplest illustration of the above general theory is provided by Example 1.1 in which there are two groups of observations with n_1 and n_2 individuals respectively. The linear logistic model (2.18) is conveniently written in the form

$$\boldsymbol{\lambda} = \begin{pmatrix} 1 \\ \vdots \\ 1 \\ \dots \\ 1 \\ \vdots \\ 1 \end{pmatrix} \alpha + \begin{pmatrix} 0 \\ \vdots \\ 0 \\ \dots \\ 1 \\ \vdots \\ 1 \end{pmatrix} \varDelta, \qquad (4.20)$$

where the vectors of n elements are partitioned into sections of n_1 and n_2 components. The matrix a of the general model is here the $n \times 2$ matrix formed from the two columns in (4.20). In the general notation α corresponds to β_1, the nuisance parameter, and $\varDelta$ to $\beta_2 = \beta$, the parameter of interest.

The sufficient statistics are formed from the scalar product of the vector of binary observations with the vectors on the right-hand side of (4.20). Thus

$$T_1 = \sum_{i=1}^{n_1+n_2} Y_i = R_1 + R_2, \qquad T_2 = \sum_{i=n_1+1}^{n_1+n_2} Y_i = R_2, \quad (4.21)$$

where R_1 and R_2 are the numbers of successes in the two groups.

In accordance with the general discussion of §4.2, inference about $\varDelta$ is based on the distribution of the random variable $T = T_2 = R_2$ given that the observed value of T_1 is t_1. That is, we need the distribution of R_2, the number of successes in the second group, given that the total number of successes in all is t_1.

To use the general formulae of §4.2(iii), we first have to calculate the combinatorial coefficients in (4.7) to (4.11). These can be found from the generating function (4.5), which in this particular problem is

$$C(\zeta_1, \zeta_2) = (1 + \zeta_1)^{n_1} (1 + \zeta_1 \zeta_2)^{n_2}. \qquad (4.22)$$

Thus $c(t_1, t_2)$, the coefficient of $\zeta_1^{t_1} \zeta_2^{t_2}$, is

$$c(t_1, t_2) = \binom{n_1}{t_1 - t_2} \binom{n_2}{t_2}, \qquad (4.23)$$

and therefore, from (4.8),

$$p_T(t; \varDelta) = \frac{\binom{n_1}{t_1 - t} \binom{n_2}{t} e^{\varDelta t}}{\sum_u \binom{n_1}{t_1 - u} \binom{n_2}{u} e^{\varDelta u}}. \qquad (4.24)$$

An important special case is obtained when $\varDelta = 0$; then (4.24) becomes the hypergeometric distribution

$$p_T(t; 0) = \frac{\binom{n_1}{t_1 - t} \binom{n_2}{t}}{\binom{n_1 + n_2}{t_1}}, \qquad (4.25)$$

since the sum in the denominator of (4.24) can be given explicitly when $\Delta = 0$.

There is a direct argument for (4.25) which is also important in a more general context (§5.4). Given t_1, i.e. the total number of 1's, and that $\Delta = 0$, all distinct binary vectors with t_1 1's and $n_1 + n_2 - t_1$ 0's, have equal probability. Hence the number of 1's in the second sample of size n_2 can be regarded as the number of 1's in a sample of size n_2 drawn randomly, without replacement, from a finite population of t_1 1's and $n_1 + n_2 - t_1$ 0's. It is well known that random sampling of this finite population without replacement leads to the hypergeometric distribution (4.25) (Feller, 1968, §II.6).

The 'exact' test of the null hypothesis $\Delta = 0$ is obtained by computing tail areas of (4.25), i.e. by substituting (4.25) into (4.15) to (4.17). This is Fisher's 'exact' test (Pearson and Hartley, 1966, Table 38).

It follows from the known properties of the hypergeometric distribution, or from results on the sampling of general finite populations, that if $E(T; 0)$ and $\text{var}(T; 0)$ denote the mean and variance of (4.25), then

$$E(T; 0) \quad = n_2 t_1 / (n_1 + n_2), \tag{4.26}$$

$$\text{var}(T; 0) = \{n_1 n_2 t_1 (n_1 + n_2 - t_1)\} / \{(n_1 + n_2)^2 (n_1 + n_2 - 1)\}. \tag{4.27}$$

An approximation to the tail areas (4.15) and (4.16) can be obtained by using the standard normal integral with argument

$$\frac{|t - E(T; 0)| - \frac{1}{2}}{\sqrt{\text{var}(T; 0)}}. \tag{4.28}$$

The statistic (4.28) differs very slightly from the square root of the usual chi-squared statistic, corrected for continuity (Pearson, 1947).

The computation of confidence limits from (4.18), (4.19) and (4.24) is in principle straightforward; in practice the simpler method of §3.3, using the empirical logistic transform, will usually be adequate. Should there be special interest in a value Δ_0 of Δ then, as noted above, if it is required that the approximate method be checked, (4.15) and (4.16) should be computed. This is probably the main practical use of the exact non-null formulae.

Gart (1962a) has compared the exact procedure with the logistic transform and other approximate procedures. The following example is based on the data of Example 1.1.

Example 4.1 *Further analysis of* 2×2 *table.* For the data of Table 1.2, the observed numbers of successes in the two groups are $r_1 = 32$ and $r_2 = 60$. The numbers of trials are $n_1 = 43$ and $n_2 = 63$. Thus the distribution (4.24) for the statistic $T = R_2$ is

$$p_T(t; \Delta) = \frac{\binom{43}{92-t}\binom{63}{t}e^{\Delta t}}{\sum_{u=49}^{63} \binom{43}{92-u}\binom{63}{u}e^{\Delta u}}.$$

The one-sided tail areas for $t = 60$ corresponding to $\Delta = \Delta_0$ are

$$P_+(60; \Delta_0) = \sum_{t=60}^{63} p_T(t; \Delta_0) \quad \text{and} \quad P_-(60; \Delta_0) = \sum_{t=49}^{60} p_T(t; \Delta_0).$$

In particular, for $\Delta_0 = 0$, corresponding to no difference between the groups, the relevant one-sided area is

$$P_+(60; 0) \simeq 0 \cdot 0025.$$

This may be compared first with the two-sided value of $0 \cdot 004$ obtained in Example 3.1 using the unweighted empirical logistic transform. Also we can compare the answer with that from a normal approximation using the exact mean and variance of T. For this we compute the statistic (4.28), namely

$$\frac{|60 - 54 \cdot 68| - \frac{1}{2}}{\sqrt{2 \cdot 957}} = 2 \cdot 80$$

corresponding to a one-sided significance level of $0 \cdot 0026$. This agrees very closely with the exact test and is in reasonably good agreement with the normal deviate of $2 \cdot 91$, obtained in Example 3.1, by the use of the empirical logistic transform.

To obtain conservative confidence limits for Δ, the simplest procedure is to compute $P_+(60; \Delta_0)$ and $P_-(60; \Delta_0)$ for a trial series of values of Δ_0 and to plot the results against Δ_0 on probability paper. Graphical interpolation will then give the values satisfying (4.18) and (4.19). A conservative 95% confidence interval for Δ obtained in this way is $(0 \cdot 473, 3 \cdot 52)$; the corresponding interval obtained in Example 3.1 by the empirical logistic transform is $(0 \cdot 592, 3 \cdot 03)$. Detailed comparison is difficult, because a method based on a normal approximation without continuity correction might be expected to give a probability of coverage averaging approximately to the nominal value, in this case 0·95, over a range of parameter values; the first interval, on the other hand, is conservative, i.e. has a probability of coverage of at least the nominal value for all parameter values. This distinction accounts for at least some of the difference between the two answers.

4.4 Agreement between a sequence of observations and a set of probabilities

As a less standard, but in some ways more elementary, example of analysis based on a logistic model we consider the testing of agreement between a series of binary observations and set of numbers which purport to be the probabilities of success. Let $Y_1, \ldots, Y_n$ be independent binary random variables and let $p_1, \ldots, p_n$ be given constants, where $0 \leqslant p_i \leqslant 1$, the hypothesis to be examined being

$$\text{prob}\,(Y_i = 1) = p_i \qquad (i = 1, \ldots, n). \tag{4.29}$$

Given a large amount of data, it would be possible to check (4.29) by forming sub-groups with nearly constant p_i; the proportion of successes in each sub-group can then be compared with the relevant p_i. This method is not applicable with relatively small amounts of data.

One method of deriving a small sample test is to consider first a family of models derived from (4.29) representing a translation from (4.29) on a logistic scale. Suppose in fact that the logistic transform for the ith trial is

$$\log\left(\frac{p_i}{1 - p_i}\right) + \beta_1,$$

so that the corresponding probabilities of success and failure are

$$\frac{p_i\, e^{\beta_1}}{p_i\, e^{\beta_1} + 1 - p_i} \quad \text{and} \quad \frac{1 - p_i}{p_i\, e^{\beta_1} + 1 - p_i}. \tag{4.30}$$

Thus the sufficient statistic is the total number of successes,

$$T_1 = \sum Y_i. \tag{4.31}$$

The probability generating function of T_1 can be written down immediately from (4.30). In particular, under the null hypothesis $\beta_1 = 0$,

$$E(T_1; \beta_1 = 0) = \sum p_i, \qquad \text{var}(T_1; \beta_1 = 0) = \sum p_i(1 - p_i). \quad (4.32)$$

The distribution of T_1 can be approximated by a normal distribution with a continuity correction; in doubtful cases the skewness and kurtosis can be calculated or exact probabilities obtained.

The statistic T_1 examines only whether the p_i are systematically too high or too low; we also need to see whether they are too clustered or too dispersed. For this, suppose that the logistic transform for the ith trial is given by another single parameter model, namely

$$\beta_2 \log\left(\frac{p_i}{1 - p_i}\right),$$

so that the probabilities of success and failure are respectively

$$\frac{p_i^{\beta_2}}{p_i^{\beta_2} + (1 - p_i)^{\beta_2}} \quad \text{and} \quad \frac{(1 - p_i)^{\beta_2}}{p_i^{\beta_2} + (1 - p_i)^{\beta_2}}. \quad (4.33)$$

The sufficient statistic for β_2 can be obtained most conveniently by scoring for the ith observation

$$\begin{cases} \log(2p_i) & \text{if } Y_i = 1, \\ \log\{2(1 - p_i)\} & \text{if } Y_i = 0, \end{cases} \quad (4.34)$$

and by defining T_2 to be the total score. The factor 2 is included for convenience and to make events of probability $\frac{1}{2}$ score zero. The cumulants of T_2 can be written down and, when $\beta_2 = 1$, the value for the null hypothesis is

$$E(T_2; \beta_2 = 1) = n \log 2 + \sum p_i \log p_i + \sum (1 - p_i) \log(1 - p_i),$$

$$\text{var}(T_2; \beta_2 = 1) = \sum p_i(1 - p_i) [\log\{p_i/(1 - p_i)\}]^2. \quad (4.35)$$

In practice we would examine both T_1 and T_2. Should we need to obtain a single test combining both statistics, the simplest procedure is to find the covariance matrix $\boldsymbol{\Omega}$ of T_1 and T_2 from the variances (4.32), and (4.35) and the covariance of T_1 and T_2 under (4.29), namely

$$\sum p_i(1 - p_i) \log\{p_i/(1 - p_i)\}.$$

Then if $\mathbf{U} = \{U_1, U_2\}$ is the row vector of differences from expectation, where $U_s = T_s - E(T_s; \beta_1 = 0, \beta_2 = 1)$, then under the null hypothesis

$$\mathbf{U}\mathbf{\Omega}^{-1}\mathbf{U}' \qquad\qquad (4.36)$$

is distributed approximately as chi-squared with two degrees of freedom.

Applications of these results in testing goodness of fit of models are outlined in §6.6; there the p_i's, instead of being given *a priori*, are estimated from the data.

One possible application of the results of this section is in the testing of conformity between subjective probabilities, p_i, and realized events. Even here, it will usually be fruitful to enquire where the subjective probabilities come from, and on what evidence they are based, rather than merely to accept them as given. If there is a discrepancy between the subjective probabilities and the observations, it may be reasonable to assume that the 'objective' probability of an event whose subjective probability is p_i has a logistic transform given by a combination of the two models above, namely

$$\beta_1 + \beta_2 \log\left(\frac{p_i}{1 - p_i}\right). \qquad\qquad (4.37)$$

The parameters β_1 and β_2 can be estimated, for example, by the methods of Chapter 6, and the resulting equation used to convert the p_i's into estimated 'objective' probabilities.

CHAPTER 5

Further problems concerning a single parameter

5.1 Introduction

We now turn to some further situations in which the 'exact' results of §4.2 can be used fruitfully to examine a single unknown parameter. Broadly, these situations are of two types:

(i) generalizations of the 2×2 contingency table (§§5.2, 5.3 and 5.6);

(ii) regression problems, in particular those analogous to simple linear regression (§§5.4 to 5.6).

In addition in §5.7 we deal briefly with a time series problem.

In most cases a relatively simple significance test can be obtained for a null hypothesis that a logistic parameter is zero, but calculation of confidence limits is more awkward computationally.

5.2 Matched pairs

Suppose that individuals are paired and that in each pair one individual is assigned at random to treatment 1, the other to treatment 2. On each individual a binary response is observed. The possible responses on a pair of individuals are thus 00, 10, 01 and 11, the response to treatment 1 being written first. Let Y_{j1} and Y_{j2} represent the responses on the two individuals forming the jth pair.

Let the numbers of pairs with the four types of response be R^{00}, R^{10}, R^{01} and R^{11}. Then $\sum R^{lm}$ is equal to the number of pairs, say k, the total number of binary responses being $n = 2k$. Paired data can, of course, arise also from observational rather than from experimental studies.

The analysis depends appreciably on the model thought to be appropriate. Thus if it were assumed provisionally that all individuals respond independently with probabilities of success θ_1 and θ_2, for treatments 1 and 2, then by arguments of sufficiency only the total numbers of successes $R^{10} + R^{11}$ and $R^{01} + R^{11}$ in the two treatment groups need be considered. They can be compared by the procedure of §4.3. Such an analysis would, however, ignore the correlation between the two individuals in a pair.

Suppose, then, to represent the pairing we adopt a linear logistic model that is analogous to the normal-theory linear model commonly used for paired comparison quantitative data. Then, for the jth pair, the logistic transforms for treatments 1 and 2 are respectively

$$\alpha_j \text{ and } \alpha_j + \Delta, \tag{5.1}$$

where α_j is a nuisance parameter characteristic of the jth pair, and Δ is a treatment effect assumed constant on the logistic scale.

In the general analysis of §4.2, the statistics associated with the nuisance parameters, and hence used for conditioning, are the pair totals $Y_{j1} + Y_{j2}$ $(j = 1,\ldots,k)$. The statistic associated with the parameter Δ is the total number of successes for treatment 2, i.e. $T = R^{01} + R^{11}$. To examine the conditional distribution of T given $Y_{j1} + Y_{j2}$ $(j = 1,\ldots,k)$, we argue as follows. Any pair for which $Y_{j1} + Y_{j2} = 0$ must contribute zero to T. Any pair for which $Y_{j1} + Y_{j2} = 2$ must contribute one to T. Hence only pairs for which $Y_{j1} + Y_{j2} = 1$, i.e. only the pairs with 'mixed' responses 01 and 10, contribute to T an amount to be regarded as random. Thus the conditional distribution required is in effect that of the number R^{01} of pairs 01, given that $R^{01} + R^{10} = m$, the total number of 'mixed' pairs.

Under (5.1) the conditional probability that the jth pair contributes one to R^{01}, given that it is 'mixed', is

$$\text{prob}\,(Y_{j1} = 0,\, Y_{j2} = 1 \,|\, Y_{j1} + Y_{j2} = 1)$$

$$= \frac{\left(\dfrac{1}{1 + e^{\alpha_j}}\right)\left(\dfrac{e^{\alpha_j + \Delta}}{1 + e^{\alpha_j + \Delta}}\right)}{\left(\dfrac{1}{1 + e^{\alpha_j}}\right)\left(\dfrac{e^{\alpha_j + \Delta}}{1 + e^{\alpha_j + \Delta}}\right) + \left(\dfrac{e^{\alpha_j}}{1 + e^{\alpha_j}}\right)\left(\dfrac{1}{1 + e^{\alpha_j + \Delta}}\right)}$$

$$= \frac{e^{\Delta}}{e^{\Delta} + 1}. \tag{5.2}$$

Because this is the same for all pairs, it follows that the conditional distribution of R^{01} is binomial with index m and parameter (5.2). In particular, under a null hypothesis $\Delta = 0$, the binomial parameter equals $\frac{1}{2}$.

It is thus possible to test hypotheses about, or to obtain confidence limits for, Δ, using methods for a single binomial sample. Further, given several independent sets of data of the above form, each with its appropriate Δ, we may set up a linear logistic model for the Δ's and then use some of the other special techniques given in this book.

Example 5.1 *Some psychiatric paired comparison data.* Maxwell (1961, p. 28) has given some data on twenty-three matched pairs of depressed patients, one member of each pair being classed as 'depersonalized', 1, the other as 'not depersonalized', 2. After treatment each patient is classified as 'recovered', coded as $Y = 1$, or 'not recovered', coded as $Y = 0$. Table 5.1 summarizes the results.

TABLE 5.1

Recovery of twenty-three pairs of patients

1 Depersonalized	2 Not depersonalized	No. of pairs
Response		
0	0	2
1	0	2
0	1	5
1	1	14

To test the difference between groups, taking as the null hypothesis $\Delta = 0$, we consider the seven 'mixed' pairs and ask whether the split into two and five is consistent with binomial sampling with seven trials and parameter $\frac{1}{2}$. Note that, in particular, the fourteen pairs in which both patients recover are disregarded. The effective numbers of observations are very small and in practice a formal test would not be necessary. The 'exact' two-sided level of significance is, however,

$$2\left\{\binom{7}{0} + \binom{7}{1} + \binom{7}{2}\right\}\frac{1}{2^7} = \frac{29}{64}.$$

5

The pairs giving responses 00 or 11 are ignored because in (5.1) each pair has an arbitrary associated parameter α_j and there is the very strong requirement that the probability properties of the method of analysis must be the same whatever the values of $\alpha_1, \ldots, \alpha_k$. Thus a pair giving a response 00 might have had a very large negative α_j disguising the presence of a treatment effect $\varDelta$. If, however, some restriction is placed on the variation of the α_j's, some relevant information may be contained in the numbers of 00 and 11 pairs. In fact the occurrence of a large number of such pairs would often be evidence that $\varDelta$ is small. A possible model of a 'random effects' type is outlined in Exercise 20. Note also that the availability of a concomitant variable for each individual or pair would allow a more economical model; see Exercise 21.

A serious criticism of the model (5.1) is that there is no check from the data on its adequacy. Note, however, that as a test of significance of $\varDelta = 0$, the binomial test has the correct probability properties whenever there is no treatment effect; the model (5.1) serves to indicate one special set of circumstances under which rejecting the pairs giving responses 00 or 11 is the efficient thing to do. An indirect justification of the model would be obtained if we had several sets of data comparing the two treatments under different conditions and with different overall success rates, the data being consistent with a single constant $\varDelta$.

5.3 Several 2 × 2 contingency tables

In the problem treated in §5.2, the observations on each pair of individuals constitute a rather degenerate form of 2×2 contingency table, in which in the notation of Table 1.1, $n_1 = n_2 = 1$, i.e. there is one observation in each group or on each treatment. A natural and important generalization of the problem of §5.2 is thus to consider a set of k general 2×2 contingency tables, all involving a comparison of the same two treatments; see Example 1.2.

Quite often the k tables will have been obtained by subdividing data initially in the form of a single 2×2 table; the dangers of not partitioning the data in this way are illustrated in extreme form in Exercise 3.

Suppose that in the jth table R_{j1} and R_{j2} are the numbers of successes in the two groups, the corresponding sample sizes being n_{j1} and n_{j2}.

Essentially two types of problem arise. The first is concerned with the possible homogeneity of the group (or treatment) differences in the separate tables. For example, are the different tables reasonably consistent with a treatment effect constant on the logistic scale? Secondly, we may tentatively assume this constant effect and wish to obtain confidence limits for it and, in particular, to test the significance of the effect, combining the evidence from the separate tables.

The first type of problem will be deferred to Chapter 6, where we deal with methods in which several parameters are of simultaneous interest. Suppose, then, that we tentatively take a model with a constant effect on the logistic scale; we allow arbitrary differences between tables in, say, the probabilities in group 1. For the jth table we take the logistic transforms of the probabilities of success in groups 1 and 2 to be respectively

$$\alpha_j \quad \text{and} \quad \alpha_j + \Delta.$$

We can now either appeal to the general results of §4.2 or proceed again from first principles to write down the combined likelihood of all observations. It is found that inference about Δ is based on the conditional distribution of

$$T = \sum R_{j2} \tag{5.3}$$

given the marginal totals of all tables.

Now the distribution of R_{j2} for a particular j is given by the generalized hypergeometric distribution (4.24) and the required distribution of T is the convolution of k of these distributions. It is clear that, except in some simple cases, notably that of §5.2, this is impracticable for exact calculations. However, we can test the null hypothesis that $\Delta = 0$ by noting that from (4.26) and (4.27) the mean and variance of T are

$$E(T;0) = \sum \frac{n_{j2} t_{j1}}{n_{j1} + n_{j2}}, \tag{5.4}$$

$$\text{var}\,(T;0) = \sum \frac{n_{j1} n_{j2}(n_{j1} + n_{j2} - t_{j1}) t_{j1}}{(n_{j1} + n_{j2})^2 (n_{j1} + n_{j2} - 1)}, \tag{5.5}$$

where t_{j1} is the observed total number of successes in the jth table. A normal approximation, with continuity correction, for the distribution of T will nearly always be adequate; the approximation is good even for a single table and will be improved by convolution. In exceptional cases a correction based on the third and fourth cumulants can be introduced.

This procedure leads to a combined test of significance from several independent 2×2 contingency tables. It is preferable to tests sometimes proposed based on the addition of chi or chi-squared values, because allowance is made for the differing amounts of information in the separate tables. However, if the tables show effects in opposite directions or if, say, only one of the tables has an effect, the use of T will not be effective. In practice these possibilities can never be excluded and it will be a wise precaution to supplement (5.4) and (5.5), for example by graphical analyses of the separate estimates of Δ based on the empirical logistic transform; see §6.3.

To estimate Δ by confidence limits the most practicable general procedure is to use the empirical logistic transform; see §§3.3 and 6.3. An estimate of Δ and its approximate standard error can be computed from each table and hence a weighted mean and its standard error obtained.

Example 1.2 is special in that $n_{j2} = 1$ for all j; it is intermediate between the matched pair problem and the general case in which n_{j1} and n_{j2} are not one. In Example 1.2 the contribution of the jth set to the combined statistic T is either 0 or 1 and under the null hypothesis the probability generating function of the contribution is

$$\frac{n_{j1} + 1 - t_{j1}}{n_{j1} + 1} + \frac{t_{j1}}{n_{j1} + 1} \zeta, \qquad (5.6)$$

where t_{j1} is the observed value of the conditioning statistic $R_{j1} + R_{j2}$ and ζ is the argument of the probability generating function. Thus the probability generating function of T under the null hypothesis is

$$\prod \left(\frac{n_{j1} + 1 - t_{j1}}{n_{j1} + 1} + \frac{t_{j1}}{n_{j1} + 1} \zeta \right). \qquad (5.7)$$

For general Δ, (5.7) is replaced by

$$\prod \{1 - \theta_j(\Delta) + \theta_j(\Delta)\zeta\}, \qquad (5.8)$$

where
$$\theta_j(\varDelta) = \frac{t_{j1}\, e^{\varDelta}}{t_{j1}\, e^{\varDelta} + n_{j1} - t_{j1} + 1}$$

is the conditional probability that $R_{j2} = 1$, given that $R_{j1} + R_{j2} = t_{j1}$.

Example 5.2 *The crying of babies*. We can apply these results to the data of Table 1.3, where the treatments 1 and 2 refer to control and experimental babies and the special condition $n_{j2} = 1$ holds, there being only one experimental baby on each day. The test statistic T, the total number of successes in group 2, has the observed value 15 and, from (5.4) and (5.5),

$$E(T;0) = 11\cdot47, \qquad \text{var}\,(T;0) = 3\cdot420.$$

Thus the standardized deviate, corrected for continuity, is $3\cdot03/\sqrt{3\cdot420} = 1\cdot64$, corresponding to significance at the 5% level in a one-sided normal test. It is fairly clear that in this example the null distribution of T will be negatively skew and hence that the normal approximation will give too large a value. As a check on this the coefficients of ζ^{18}, ζ^{17}, ζ^{16} and ζ^{15} in the generating function (5.7) were evaluated recursively by computer. The 'exact' significance level, the sum of these coefficients, is $0\cdot045$.

Cox (1966a) has considered this and other ways in which these data might be analyzed.

5.4 Linear regression

Suppose now that to each individual is attached a single regressor variable x thought to influence the probability of success. The data are represented by n pairs (x_1, Y_1), ..., (x_n, Y_n), where the Y_i's are assumed to be independent binary random variables. Examples 1.4 to 1.6 illustrate this situation. There are two particular cases of special interest. The first has $x_i = i$ covering, in particular, the dependence of probability of success on the serial order of the trial (Example 1.4). In the second the x_i's are grouped into a fairly small number, g, of sets, each with a constant value of the regressor variable. This can be regarded as a $2 \times g$ contingency table (Example 1.6) in which a 'score' is attached to each column.

As in other regression problems graphical methods are very important. We can, for example, group the x-values, if they are not grouped to begin with, and plot the empirical logistic transform, (3.13), against x.

Suppose that we analyze the data in the light of the linear logistic model,

$$\lambda = \alpha \begin{pmatrix} 1 \\ \vdots \\ 1 \end{pmatrix} + \beta \begin{pmatrix} x_1 \\ \vdots \\ x_n \end{pmatrix} ; \qquad (5.9)$$

in the problems we shall consider here, β is the parameter of primary interest. Note, however, that in the application to stimulus–response curves the model (5.9) may hold but interest will not normally be primarily in β; it may, for example, be in the stimulus, $-\alpha/\beta$, at which the probability of success is $\frac{1}{2}$.

The general results of §4.2 take on a particularly simple form here. The conditioning statistic is $\sum Y_i$, the total number of successes, its observed value being, say, t_1. The statistic associated with the parameter of interest is $\sum x_i Y_i$, i.e. the sum of the x_i's over those individuals giving a success. Given t_1, the statistic $T = \sum x_i Y_i$ is the total of a sample of size t_1 drawn from the finite population $\mathcal{X} = \{x_1, \ldots, x_n\}$. Further, in the special case when $\beta = 0$, so that the probability of success is constant, it is clear that all distinct samples of size t_1 have equal probability, i.e. that T is the total of a sample drawn randomly without replacement from the finite population $\mathcal{X}$.

Now the cumulants and moments of the null distribution of T are known; see, for example, Kendall and Stuart (1963, pp. 300–304). In fact,

$$E(T;0) = t_1 m_1,$$

$$\mathrm{var}\,(T;0) = \frac{t_1 (n - t_1) m_2}{(n - 1)},$$

$$\mu_3(T;0) = \frac{t_1 (n - t_1)(n - 2t_1) m_3}{(n - 1)(n - 2)}, \qquad (5.10)$$

$$\mu_4(T;0) = \frac{t_1 (n - t_1)}{(n - 1)(n - 2)(n - 3)} \{(n^2 - 6nt_1 + n + 6t_1^2) m_4$$

$$+ 3(t_1 - 1) n(n - t_1 - 1) m_2^2\},$$

where $m_1 = \sum x_i / n$, $m_s = \sum (x_i - m_1)^s / n$ $(s = 2, 3, \ldots)$.

Here the zeros in $E(T;0),\ldots,$ indicate that $\beta = 0$. The dependence on the condition $T_1 = t_1$ is not shown on the left-hand side.

A normal approximation to the null distribution of T will often be adequate, but it can be checked by computing the skewness and kurtosis of T and, if necessary, corrections for non-normality can be introduced. The use of a normal approximation is supported by limit theorems which give under weak conditions the asymptotic normality of T, as t_1 and n tend to infinity with t_1/n fixed.

To get some insight into the rate of approach to normality, it is useful to write $f = t_1/n$ for the effective sampling fraction and to take n large with f fixed. Then

$$E(T;0) = t_1 m_1,$$

$$\operatorname{var}(T;0) = t_1 m_2(1-f)\left\{1 + \frac{1}{n} + o\left(\frac{1}{n}\right)\right\}, \qquad (5.11)$$

$$\gamma_1(T;0) \sim \frac{(1-2f)\gamma_1}{\sqrt{\{nf(1-f)\}}},$$

$$\gamma_2(T;0) \sim -\frac{6}{n} + \frac{\{1 - 6f(1-f)\}\gamma_2}{nf(1-f)},$$

where $\gamma_1 = m_3/m_2^{3/2}$ and $\gamma_2 = m_4/m_2^2 - 3$ refer to the finite population, $\mathscr{X}$.

The special case when $\mathscr{X}$ consists only of 0's and 1's corresponds to the 2×2 contingency table (§4.3), and (5.10) and (5.11) then refer to the hypergeometric distribution. In other particular cases it may be possible to find the distribution of T explicitly; by (4.7) this requires in effect the calculation of $c(t_1,t)$, the number of distinct samples of size t_1 from $\mathscr{X}$ that give a sample total of t. Once this is found, the null and non-null distributions of T follow from (4.8).

One way of approximating to the distribution of T for small non-zero β is by expansion of the general relation (4.13) for cumulant generating functions. In particular,

$$E(T;\beta) = E(T;0) + \beta \operatorname{var}(T;0) + \tfrac{1}{2}\beta^2 \mu_3(T;0)$$
$$+ \tfrac{1}{6}\beta^3[\mu_4(T;0) - 3\{\operatorname{var}(T;0)\}^2] + \cdots, \quad (5.12)$$

with analogous expressions for the higher cumulants.

There are two special cases of particular interest, corresponding respectively to tests for serial order effects (Example 1.4) and to the analysis of a $2 \times g$ contingency table for trend (Example 1.6). In the first of these problems the finite population $\mathscr{X}$ is $\{1, 2, \ldots, n\}$. The test statistic is, under the null hypothesis, the total of a random sample of size t_1 drawn without replacement from $\mathscr{X}$. This leads to the same null probability distribution as arises in the two-sample Wilcoxon test (Kendall and Stuart, 1967, p. 492), and tables of its exact distribution (Fix and Hodges, 1955), can be used if t_1 or $n - t_1$ is small. Note that in the Wilcoxon test the 0's and 1's correspond to a coding of the two samples and the values 1, 2, $\ldots$, n are the rank numbers of the pooled data; the roles of the random and non-random variables in that problem are thus complementary to those in the regression problem.

A simple, if artificial, numerical example illustrates some of the above formulae.

Example 5.3 A series of seven individuals. The moments of the finite population $\{1, 2, \ldots, 7\}$ are

$$m_1 = \tfrac{1}{7} \sum_{i=1}^{7} i = 4, \quad m_2 = \tfrac{1}{7} \sum_{i=1}^{7} (i-4)^2 = 4, \quad m_3 = 0, \quad m_4 = 28.$$

If there are three successes in the seven trials, we are concerned with the sum of a sample of size $t_1 = 3$ from this finite population. Hence, from (5.10),

$$E(T; 0) = 12, \qquad \mathrm{var}\,(T; 0) = 8,$$

$$\mu_3(T; 0) = 0, \qquad \mu_4(T; 0) = 156 \cdot 8,$$

so that

$$\gamma_1(T; 0) = 0, \qquad \gamma_2(T; 0) = -11/20.$$

Table 5.2 gives the upper tail of the null distribution of T obtained by enumeration of the thirty-five distinct samples of size three from the finite population, or by use of a normal approximation with continuity correction, or by use of an adjustment based on the Edgeworth expansion. The effect of the correction for kurtosis is small. An alternative way of allowing for non-normality is by Table 42 of Pearson and Hartley (1966), based on the fitting of Pearson curves.

TABLE 5.2

Comparison of exact and approximate tail areas

Observed value t	$\text{prob}(T \geqslant t; \beta = 0)$		
	Exact	Normal approximation	Edgeworth approximation
18	$1/35 = 0\cdot029$	$0\cdot026$	$0\cdot024$
17	$2/35 = 0\cdot057$	$0\cdot056$	$0\cdot057$
16	$4/35 = 0\cdot114$	$0\cdot108$	$0\cdot116$
15	$7/35 = 0\cdot200$	$0\cdot188$	$0\cdot200$
14	$11/35 = 0\cdot314$	$0\cdot298$	$0\cdot309$
13	$15/35 = 0\cdot429$	$0\cdot430$	$0\cdot435$

A second illustrative example could be based on the 2×3 contingency table, Table 1.7, where the finite population would be, in an obvious notation,

$$\{-1^{497}, 0^{560}, 1^{293}\}.$$

For null hypotheses representing zero effects, the tests of this section have an important non-parametric property. The logistic model is essentially used to obtain an appropriate test statistic; however the null hypothesis and the distribution used to obtain a significance level hold very generally.

5.5 Combination of regressions

The discussion of §5.3 concerns several 2×2 contingency tables in which it is assumed provisionally that the logistic treatment effect is the same in all tables, even though the individual probabilities of success change from table to table. A closely analogous discussion holds when there are k independent sets of data, within each of which, regression on an independent variable x is to be considered. Suppose that the logistic regression equation has the form

$$\alpha^{(j)} + \beta x_{ij} \qquad (i = 1, \ldots, n_j; j = 1, \ldots, k), \qquad (5.13)$$

where x_{ij} is the value of the independent variable for the ith individual in the jth set, $\alpha^{(j)}$ is a parameter for the jth set and β is the common regression coefficient on x. We consider in Chapter 6 the problem of examining the consistency of the data with such a model.

Under (5.13), the conditioning statistics are the separate sample numbers of successes $t_1^{(1)}$, ..., $t_1^{(k)}$ in the k sets, and the statistic associated with β is

$$T = \sum T^{(j)}, \tag{5.14}$$

where $T^{(j)}$ is the test statistic associated with the jth set. Since the different sets are independent, the distribution of T, for any value of β, is the convolution of k distributions of the type investigated in §5.4. The adequacy of a normal approximation will usually be improved by convolution.

If k is large it will be sensible to make a graphical analysis of the separate $T^{(j)}$'s. For instance, they can be separately standardized to have zero mean and unit variance and then plotted against relevant properties of the groups. Also they can be ranked and plotted on probability paper.

As a special case we can deal with some problems of significance testing connected with regression on two variables, say z and x. If for the ith individual the values of the independent variables are z_i and x_i, a natural logistic model generalizing the simple regression model is

$$\lambda_i = \alpha + \beta_1 z_i + \beta_2 x_i. \tag{5.15}$$

Unfortunately, if we are now interested in, say, $\beta_2 \equiv \beta$, it is not usually simple to find the appropriate conditional distribution. The following approach (Hitchcock, 1966) will often be reasonable instead.

Group the z_i's into a fairly small number k of sets; of course the z_i's may already be so grouped. Then consider, instead of (5.15), the model in which for an individual in the jth set, i.e. having the jth value of z, with a value x_{ij} for x, the logistic transform is

$$\alpha^{(j)} + \beta x_{ij}. \tag{5.16}$$

This is in one sense more general than (5.15), in that it does not postulate a linear regression on z; on the other hand, there may be some artificiality in grouping the values of z.

If the model (5.16) is a reasonable basis for the investigation of the partial regression on x, the analysis is based essentially on the discussion of §5.4. The same argument could be applied, for example, with three regressor variables z, u and x, provided that it is feasible to group the (z, u) combinations into a fairly small number of sets.

The procedure is illustrated by the following example.

Example 5.4 *Regression on two variables.* Table 5.3a gives the number, r, of ingots not ready for rolling, out of n tested, for a number of combinations of the regressor variables, heating time and soaking time. Because the total number of ingots not ready for rolling is small, the information about the regression is likely to be slight. The tests based on (5.14) are, however, useful in a preliminary analysis of the data.

First consider possible regression on heating time, u_1. For each row of the table, i.e. for each fixed value of soaking time, u_2, we compute the test statistic for regression on u_1 and its null mean and variance. These are shown in Table 5.3b. Thus in the first row the test statistic is

$$1 \times 27 + 3 \times 51 = 180,$$

and its null mean and variance are given by (5.10) with $t_1 = 4$, $n = 110$ and with m_1 and m_2 being the moments of the finite population of 110 values of u_1; in fact, $m_1 = 24 \cdot 35$ and $m_2 = 145 \cdot 07$. A similar calculation is made for the other rows and the combined test statistic, by (5.14), is simply the sum of the separate test statistics, with null mean and variance given by the sum of the separate means and variances.

With more extensive data it would be useful to plot the standardized deviates in various ways to look for systematic features; thus these values might be plotted against the values of u_2 characterizing the rows and also the ranked deviates plotted on normal probability paper. In the present case, the general conclusion is clear. In four rows out of five, the test statistic exceeds its null expectation and the pooled test statistic gives highly significant evidence of an increase, with u_1, in the probability of not being ready for rolling.

The corresponding analysis for regression on u_2 shows no evidence of departures from the null hypothesis.

To summarize the data, the most informative quantities are probably the proportions not ready for rolling at each value of u_1, pooled over rows, i.e.

$$0/55 = 0, \quad 2/157 = 0 \cdot 013, \quad 7/159 = 0 \cdot 044, \quad 3/16 = 0 \cdot 19.$$

TABLE 5.3

Number, r, of ingots not ready for rolling out of n tested

a *The data*

(First figure in each cell is r, second n)

Soaking time, u_2	Heating time, u_1				
	7	14	27	51	Total
1·0	0, 10	0, 31	1, 56	3, 13	4, 110
1.7	0, 17	0, 43	4, 44	0, 1	4, 105
2·2	0, 7	2, 33	0, 21	0, 1	2, 62
2·8	0, 12	0, 31	1, 22	0, 0	1, 65
4·0	0, 9	0, 19	1, 16	0, 1	1, 45
Total	0, 55	2, 157	7, 159	3, 16	12, 387

b *Regression on u_1*

Row	Test statistic	Null mean	Null st. dev.	Standardized deviate
1	180	97·4	23·8	3·48
2	108	74·7	16·5	2·02
3	28	36·4	11·7	−0·72
4	27	17·1	7·53	1·31
5	27	18·0	9·16	0·98
Pooled	370	243·7	33·4	3·79

c *Regression on u_2*

Col	Test statistic	Null mean	Null st. dev.	Standardized deviate
1	—	—	—	—
2	4·4	4·32	1·27	0·06
3	14·6	13·32	2·41	0·53
4	3·0	3·92	1·24	−0·74
Pooled	22·0	21·56	2·99	−0·14

These vary smoothly with u_1. A logistic regression equation could be fitted and this might be fruitful if, for example, several sets of data were involved and it was required to compare the dependences on u_1 in the different sets. With the present data, however, such fitting would normally add little information to that given directly by the above set of proportions.

5.6 Adjustment for a concomitant variable in a 2 × 2 table

Suppose that it is required to compare the probabilities of success for two groups of individuals and that for each individual a concomitant variable z is available, suspected of affecting the probability of success. We may require to compare the groups, adjusting for any differences associated with the concomitant variable. The situation is analogous to that treated by analysis of covariance in normal theory and is subject to the same broad conditions for its validity.

Sometimes, for instance in certain medical applications, an estimate of the average difference in probabilities of success may be required, the average being taken over a pre-assigned distribution of the concomitant variable, z. Thus z may be age, and probabilities of death may be standardized with respect to a 'standard' age distribution. Then the simplest procedure, at least with extensive data, is to divide the data into sets on the basis of a grouping of z. An estimated difference between groups can then be found for each set, and the weighted mean difference and its standard error found.

If an analysis in terms of a constant difference on a logistic scale is attempted, two approaches are possible, corresponding in fact to the two approaches to the regression problem contrasted in (5.15) and (5.16). If the values of z are grouped, we have the problem of §5.3, namely the analysis of replicated 2 × 2 tables. This corresponds also to the analysis of (5.16). An alternative is also to assume logistic regression on z, the simplest assumption being to have the same slope in the two groups. This leads to the model

$$\lambda_i = \begin{cases} \alpha + \beta z_i & \text{in group A,} \\ \alpha + \beta z_i + \varDelta & \text{in group B,} \end{cases}$$

exactly corresponding to the standard assumptions of linearity and parallelism involved in the simplest form of analysis of covariance.

The general procedures of this chapter now lead to an analysis for β, in particular to a simple test of the null hypothesis $\beta = 0$, and to a not so simple analysis for Δ, the parameter of main interest. The advantages and difficulties of this approach are illustrated by the following simple artificial example.

Example 5.5 Trend in a 2×2 contingency table. Consider an experiment to compare two treatments A and B which are tested in serial order in time. Table 5.4 gives some artificial data from such an experiment .

First, a test of a trend with serial order is a direct application of the results of §5.5. We calculate separately for A and B the test statistics for trend, together with their null means and variances. The test statistics are

A: 195, with null expectation 173·0 and variance 192·1,
B: 154, with null expectation 122·2 and variance 354·4.

For example, the statistic for A is the sum of the serial numbers of the successes in that group, $4 + 8 + \ldots + 28$. This is regarded as a random sample of size 11 from the finite population $\{2, 4, 7, \ldots, 28\}$.

The pooled test statistic is thus 349 leading to a standardized normal deviate, with continuity correction, of

$$\frac{|349 - 295 \cdot 2| - \frac{1}{2}}{\sqrt{546 \cdot 5}} = 2 \cdot 29.$$

For both treatments the test statistic exceeds its null expectation and the pooled statistic, significant at about the 2% level, indicates quite strong evidence that the trend apparent from the inspection of the data is unlikely to be spurious.

The formulation of a corresponding procedure for the estimation of the difference between treatments, allowing for regression on serial number, is more complicated. The parameter of interest is now Δ and the conditioning statistics are the total number t_1 of successes, where in fact $t_1 = 19$; and the sum t_2 of the serial numbers over the successes, where in fact $t_2 = 349$. The test statistic is the number t_3 of successes under treatment B. In principle we thus require to find $c(t_1, t_2, t_3)$, the number of distinct binary sequences with the relevant values of t_1, t_2 and t_3; in fact

TABLE 5.4

Comparison of two treatments tested in serial order

Serial no., i	Treatment	Response	$\hat{\theta}_i$	d_i
1	B	0	0·177	−0·46
2	A	0	0·387	−0·80
3	B	1	0·215	1·91
4	A	1	0·446	1·11
5	B	0	0·258	−0·59
6	B	0	0·282	−0·63
7	A	0	0·537	−1·08
8	A	1	0·567	0·87
9	B	0	0·361	−0·75
10	B	1	0·390	1·25
11	A	1	0·653	0·73
12	B	0	0·449	−0·90
13	B	1	0·479	1·04
14	A	1	0·730	0·61
15	A	0	0·753	−1·75
16	A	1	0·775	0·54
17	A	0	0·795	−1·97
18	B	0	0·627	−1·30
19	B	1	0·655	0·73
20	A	1	0·848	0·42
21	A	1	0·863	0·40
22	A	1	0·877	0·37
23	B	1	0·755	0·57
24	B	0	0·777	−1·87
25	A	1	0·911	0·31
26	A	1	0·920	0·29
27	B	1	0·833	0·45
28	A	1	0·936	0·26
29	B	1	0·864	0·40
30	B	1	0·878	0·37

The fitted probabilities, $\hat{\theta}_i$, and the residuals, d_i, are discussed in §§6.5 and 6.6.

$t_1 = 19$, $t_2 = 349$ and also t_3 varies over its full range of possible values, namely 4, …, 15.

Direct enumeration of these coefficients would be a formidable task and while it is possible to obtain an approximation by considering the asymptotic trivariate normal distribution of the three statistics, it seems simplest to make the further analysis by the general approximate techniques of Chapter 6. Therefore we leave further discussion until then.

5.7 A time series problem

As a more complicated example of a problem in which there is a single parameter of primary interest, we now consider briefly a time series taking at each point one of only two possible values. The simplest model for such a series is the two-state Markov chain in which the matrix of one-step transition probabilities is, say,

$$\begin{pmatrix} \pi_{00} & \pi_{01} \\ \pi_{10} & \pi_{11} \end{pmatrix}, \tag{5.17}$$

where $\pi_{ab} = \text{prob}\,(Y_{l+1} = b \,|\, Y_l = a)$ with

$$\pi_{00} + \pi_{01} = \pi_{10} + \pi_{11} = 1.$$

The process is completely random if and only if $\pi_{01} = \pi_{11}$.

There are essentially two independent parameters in the model and the statistical problems that occur are closely connected with those arising in the comparison of two probabilities from independent trials.

To obtain the likelihood in a convenient form, it is useful to reparametrize (5.17), writing

$$\pi_{01} = \frac{e^{\alpha}}{1 + e^{\alpha}}, \qquad \pi_{11} = \frac{e^{\alpha+\Delta}}{1 + e^{\alpha+\Delta}}. \tag{5.18}$$

Of course, no additional assumption about the process is involved in this, although there is the presumption that Δ is a suitable parameter for describing the non-randomness of the process. If and only if $\Delta = 0$, the model represents independent trials with constant probability of success.

As with other time series problems, several forms of likelihood can be considered. In the first it is assumed that the initial state is governed by the equilibrium distribution of the chain; that is, it can in effect be supposed that the initial observation is obtained after the system has been running for some time. In a second form of the likelihood the initial state is taken as a given constant. The likelihood is then a product of π_{ij}'s and can be written

$$\pi_{00}^{f_{00}} \pi_{01}^{f_{01}} \pi_{10}^{f_{10}} \pi_{11}^{f_{11}}, \tag{5.19}$$

where the so-called transition count f_{ij} is the number of times in the whole sequence that i is followed by j.

In the usual notation, let f_i . and $f._j$ denote the row and column sums of the matrix $\{f_{ij}\}$. For instance, $f._0 = f_{00} + f_{10}$ is the total number of times 0 occurs as the second member of a pair. This is either equal to or is one less than the total number of 0's in the sequence, depending on whether the initial observation is a 1 or a 0. In fact, if u and v denote the initial and final observations of the sequence of n observations and if r is the total number of 1's, then

$$f._0 = n - r - \delta_{u0}, \quad f._1 = r - \delta_{u1},$$
$$f_0. = n - r - \delta_{v0}, \quad f_1. = r - \delta_{v1},$$

(5.20)

where $\delta_{ab} = 1(a = b)$ and $\delta_{ab} = 0(a \neq b)$. A further relation closely connected with (5.20) is that if w is the number of runs of 0's and 1's in the sequence, then

$$w = 2r - 2f_{11} + (1 - u - v).$$

(5.21)

For example, in the sequence

$$1\ 0\ 1\ 0\ 1\ 0\ 1\ 1\ 0\ 1\ 1,$$

we have that $n = 11$, $r = 7$, $f_{00} = 0$, $f_{01} = 4$, $f_{10} = 4$, $f_{11} = 2$, $u = v = 1$ and $w = 9$, and (5.20) and (5.21) are easily verified.

If now we introduce into the likelihood (5.19) the logistic representation (5.18), we obtain a likelihood

$$\frac{\exp(\alpha f._1 + \Delta f_{11})}{(1 + e^\alpha)^{f_0.}(1 + e^{\alpha + \Delta})^{f_1.}}.$$

(5.22)

This is not exactly of the form required for the application of the theory of Chapter 4. In fact, the sufficient statistics are in effect f_{11}, the number of times 1 is followed by 1, r, the total number of 1's, and v, the final observation. Note that the initial observation, u, is, throughout, regarded as fixed.

Tests of the null hypothesis, $\Delta = 0$, of complete randomness closely related to the likelihood (5.22) have been widely considered; see, for example, Stevens (1939), Wald and Wolfowitz (1940) and Billingsley (1961). The most common test is based on the number of runs taken conditionally on the total number of successes, i.e. the null distribution is obtained by randomly permuting r 1's and $(n - r)$ 0's.

There are, however, several ways of proceeding formally from (5.22). The simplest is probably to argue conditionally on the

6

outcomes of the first and last trial, as well as on the total number of 1's; this leads to a distribution not depending on α but its consideration presumably involves a slight loss of information, because the sufficient statistic for the nuisance parameter is not complete. If we write T for the random variable corresponding to f_{11}, we have for its conditional distribution, given u, v and r,

$$p_T(t;\Delta) = \frac{c_{uv}(r,n-r,t)\,e^{\Delta t}}{\sum_s c_{uv}(r,n-r,s)\,e^{\Delta s}}, \qquad (5.23)$$

where the combinatorial coefficient $c_{uv}(r,n-r,t)$ is the number of distinct binary sequences of r 1's and $(n-r)$ 0's, starting with u and ending with v, and such that $f_{11}=t$. In particular, a test of the null hypothesis $\Delta = 0$ is easily obtained once the combinatorial coefficients are known.

A special case of a combinatorial result of Whittle (1955) shows that the number of binary sequences starting with u and ending with v and having transition count $\{f_{ij}\}$ is

$$\frac{f_0.\,!\,f_1.\,!}{f_{00}!\,f_{01}!\,f_{10}!\,f_{11}!} \times \frac{f_{1-u,v}}{f_{1-v}.}. \qquad (5.24)$$

Note that in (5.24) all factors are functions of $f_{11}=t$ and the conditioning variables. Further the factors that are marginal totals cancel from the numerator and the denominator of (5.23).

The null mean and variance of T can be found from (5.23) and (5.24) or, alternatively, more directly by writing

$$I_s = \begin{cases} 1 & (Y_s = Y_{s-1} = 1), \\ 0 & \text{otherwise}, \end{cases}$$

and noting that $T = \sum I_s$. Hence, once $E(I_s)$, $\mathrm{var}(I_s)$ and $\mathrm{cov}(I_{s_1}, I_{s_2})$ are calculated, $E(T)$ and $\mathrm{var}(T)$ can easily be found. It can thereby be shown that

$$E(T;0) = \frac{r^*(r-1)}{(n-2)},$$

$$\mathrm{var}(T;0) = \frac{r^*(r^*-1)(n-2-r^*)(n-1-r^*)}{(n-2)^2(n-3)}$$

$$+ (u+v)\frac{r^*(n-2-r^*)(n-1-2r^*)}{(n-2)^2(n-3)}$$

$$- \frac{2uvr^*(n-2-r^*)}{(n-2)^2(n-3)}, \qquad (5.25)$$

where $r^* = r - u - v$ is the number of 1's not at the beginning or end of the sequence.

The theory sketched above can be extended in various ways. For example, if a combined analysis of several independent sets of data is required, assuming a common value of Δ, then an argument parallel to that of §5.5 suggests a test statistic obtained by summing the separate T's. The null mean and variance are obtained from (5.25).

Again, there is a close connection between the distribution defined by (5.23) and (5.24) and that associated with the 2×2 contingency table; the distinction lies in the second factor $f_{1-u,v}/f_{1-v,.}$ in the combinatorial coefficient (5.24). Now $f_{1-v,.}$ is constant in both numerator and denominator of (5.23), so that the additional factor is either f_{10} or f_{01}, according as $v = 0$ or 1. Now in large samples, we are interested in values of the transition counts f_{ij}, near to their expectations. In particular, in testing the null hypothesis $\Delta = 0$, we are concerned with values within a few standard errors of the expectation. Over this range, f_{10} and f_{01} vary relatively slowly. Careful analysis shows that in the large-sample approximation we can treat $\{f_{ij}\}$ as the entries in an ordinary 2×2 contingency table and apply techniques for the comparison of two independent samples.

The main practical application of this analysis is that the empirical logistic transform is approximately applicable, thus justifying treating the transition counts as an ordinary contingency table. Other time series models involving logistic models are outlined in Exercises 30 and 31.

CHAPTER 6

More complex problems

6.1 Introduction

In the previous two chapters problems have been studied in which, at least in theory, an explicit 'exact' solution is possible. Here 'exact' is a technical term meaning that a probability distribution can be found depending only on the single parameter of interest and from which, for example, a theoretical significance level can be calculated without mathematical approximation. While it is, of course, a good thing to be able to make such a calculation, too much importance should not be attached to it. Any such calculation depends on an inevitably idealized model of the real situation; also the way significance levels are used in practice is sufficiently vague for great refinement in the calculations to be quite unnecessary.

Broadly speaking, the results of the previous two chapters give useful significance tests of null hypotheses of zero effects, but computation of confidence limits will usually involve extensive, if straightforward, computation. A test of a null hypothesis of an effect of a specified non-zero magnitude is a little simpler than a confidence limit calculation, since only one value of the parameter has to be examined.

We now turn to methods which rely for their justification on asymptotic theory and which, therefore, in practice involve some approximation in the distributional calculations. The usefulness of such methods is two-fold. An 'exact' solution may be available but be too complex computationally, or the relevant conditional distribution may be degenerate, or nearly so. Further, the problem may be outside the scope of the previous work; for example it may concern a null hypothesis about several parameters simultaneously.

Two types of procedure will be discussed. One is based on the empirical logistic transform introduced in Chapter 3. It has the advantage that the calculations are based on non-iterative weighted least squares and are thus standard in character. A further extremely important point is that various graphical analyses are possible with the logistic transforms, treating them as approximately normally distributed with known variance. One disadvantage of such methods is that they require the individual responses to be grouped into sets with a constant probability of success within each set; the data may initially be in this form, but if not, some arbitrary grouping will be necessary. In particular, in the multiple regression problem with many regressor variables, it will rarely be possible to achieve such grouping satisfactorily. A second disadvantage is that the analysis is not in general based on the sufficient statistics for the problem and that some loss of efficiency is thus inevitable.

The second type of procedure depends on large-sample maximum likelihood theory. This does use the sufficient statistics for the problem, but the calculations are more complex and for general use an appropriate computer program is essential. Also simple graphical methods are not so obviously available.

It would be good to give a careful statement of when these asymptotic results can safely be used. Unfortunately, this is not possible at the moment. It is probably safe to use the empirical logistic transform provided that none or very few of the groups have zero successes or zero failures. One warning against using the usual formulae for the asymptotic standard error of a maximum likelihood estimate is obtained when the log likelihood is not well approximated by a quadratic function of the parameters in the region of importance. If a specific check is required for a particular application this can be based on a simulation study under conditions closely corresponding to the data under analysis.

A quite important general point is that these methods are based on models in which the only random element is the binomial variability about the hypothetical probability of success. Sometimes, however, there will be additional components of variability. If these can be reasonably approximated by an inflation of the binomial variance by a factor $1 + \gamma$, the methods of this chapter can be applied. Certain refinements in the definition of the empirical logistic transform will no longer be appropriate, but otherwise the only change is that all variances and expected sums of squares become multiplied by $1 + \gamma$. Some forms of negative correlation, leading to under-dispersion relative to the binomial distribution, are represented by negative values of γ. We shall, however, refer to the situation discussed in this paragraph as one with inflated variance, since the case $\gamma > 0$ is likely to be the relatively more common one.

It is likely that appreciable differences will occur between the point estimates given by maximum likelihood and by the empirical logistic transform only when the model is a bad fit.

6.2 Empirical logistic transform; general

The basic properties of the empirical logistic transform have been set out in Chapter 3. There are two slightly different definitions and it is convenient first to review these. In the first definition, given by (3.13) and (3.14),

$$Z_j = \log\left(\frac{R_j + \frac{1}{2}}{n_j - R_j + \frac{1}{2}}\right), \qquad V_j = \frac{(n_j + 1)(n_j + 2)}{n_j(R_j + 1)(n_j - R_j + 1)}, \quad (6.1)$$

where R_j is the number of successes in n_j trials and V_j is an estimate of the variance associated with the transform Z_j. This is for use when interest lies in linear combinations of the Z_j's with fixed coefficients.

A second definition in effect given by (3.25) and (3.26) is used when the reciprocals of the V's are used as weights in a weighted least squares analysis. We define, for $n_j > 1$,

$$Z_j^{(w)} = \log\left(\frac{R_j - \frac{1}{2}}{n_j - R_j - \frac{1}{2}}\right), \qquad V_j^{(w)} = \frac{(n_j - 1)}{R_j(n_j - R_j)}. \quad (6.2)$$

If the linear logistic model specifies that in the jth group

$$\lambda^{(j)} = \sum_{s=1}^{p} c_{js}\beta_s,$$

then if

$$B_{st} = \sum c_{js}c_{jt}/V_j^{(w)}, \qquad U_s = \sum c_{js}Z_j^{(w)}/V_j^{(w)}, \quad (6.3)$$

the estimating equations are

$$\sum B_{st}\tilde{\beta}_t^{(w)} = U_s. \quad (6.4)$$

Zero contributions to B_{st} and U_s are taken from any group for which $R_j = 0$ or n_j.

The asymptotic distribution of the weighted estimates $\tilde{\beta}^{(w)}$ is multivariate normal with covariance matrix $\mathbf{B}^{-1}$ and the residual weighted sum of squares corresponds to unit theoretical variance. The covariance matrix and expected sum of squares are multiplied by $1 + \gamma$ under the model with inflated variance.

The refinements in (6.1) and (6.2), i.e. the addition and subtraction of constants such as $\frac{1}{2}$, are to be thought of as relatively unimportant. If, for example, the difference between (6.1) and (6.2) were to make a drastic difference to the conclusions, the whole approach would be suspect.

With the above results, all the techniques, numerical and graphical, associated with normal-theory linear models become available. It would be pointless to try and list them all and there follow merely a few illustrative examples.

6.3 Empirical logistic transform; application

Many of the advantages and disadvantages of the use of the empirical logistic transform can be brought out by discussing again the problem of §5.3 in which there are k independent 2×2 contingency tables, all comparing the same two treatments. The analysis of a single table has already been dealt with in §3.3 and involves the unweighted transform (6.1) leading for the jth table to an estimate, $\mathit{\Delta}_j$, of the logistic difference, together with an approximate variance. In the notation of §5.3,

$$\mathit{\Delta}_j = \log\left(\frac{R_{j2} + \frac{1}{2}}{n_{j2} - R_{j2} + \frac{1}{2}}\right) - \log\left(\frac{R_{j1} + \frac{1}{2}}{n_{j1} - R_{j1} + \frac{1}{2}}\right), \qquad (6.5)$$

$$V_{\mathit{\Delta}_j} = \frac{(n_{j2} + 1)(n_{j2} + 2)}{n_{j2}(R_{j2} + 1)(n_{j2} - R_{j2} + 1)}$$
$$+ \frac{(n_{j1} + 1)(n_{j1} + 2)}{n_{j1}(R_{j1} + 1)(n_{j1} - R_{j1} + 1)}. \qquad (6.6)$$

If, however, we are to combine the information from the separate tables by a weighted mean, the formulae for the weighted transformation are relevant. For the jth table we consider

$$\mathit{\Delta}_j^{(w)} = \log\left(\frac{R_{j2} - \frac{1}{2}}{n_{j2} - R_{j2} - \frac{1}{2}}\right) - \log\left(\frac{R_{j1} - \frac{1}{2}}{n_{j1} - R_{j1} - \frac{1}{2}}\right) \qquad (6.7)$$

with an associated variance

$$V_{\mathit{\Delta}_j}^{(w)} = \frac{n_{j2} - 1}{R_{j2}(n_{j2} - R_{j2})} + \frac{n_{j1} - 1}{R_{j1}(n_{j1} - R_{j1})}. \qquad (6.8)$$

Under the model in which the logistic effect Δ is the same for all k tables, the weighted least squares estimate of Δ is the weighted mean of the separate estimates (6.7), i.e. is

$$\hat{\Delta}^{(w)} = \{\textstyle\sum \hat{\Delta}_j^{(w)}/V_{\Delta_j}^{(w)}\}/\{\textstyle\sum 1/V_{\Delta_j}^{(w)}\} \qquad (6.9)$$

with approximate variance

$$\{\textstyle\sum 1/V_{\Delta_j}^{(w)}\}^{-1}. \qquad (6.10)$$

The estimate $\hat{\Delta}^{(w)}$ is not a function of the sufficient statistic for the problem, as can easily be seen by writing out explicitly the formula for $k = 2$.

If the population logistic effect Δ_j is the same for all studies, the residuals

$$\{\hat{\Delta}_j^{(w)} - \hat{\Delta}^{(w)}\}/\sqrt{V_{\Delta_j}^{(w)}} \qquad (6.11)$$

should behave approximately like the residuals for a random sample from the unit normal distribution. Thus the sum of squares of residuals will be distributed approximately as chi-squared with $k - 1$ degrees of freedom. Further the ranked values can be plotted against the expected order statistics for samples of size k from the standard normal distribution (Pearson and Hartley, 1966, Table 20).

Example 6.1 *Association between smoking and lung cancer.* Cornfield (1956) reproduced and analyzed the data in Table 6.1, given originally by Dorn (1954). Cornfield's analysis was based directly on the likelihood of the observations as a function of the differences $\Delta_1, \ldots, \Delta_k$; here we use the empirical logistic transform.

Table 6.2 gives estimates of the logistic difference for each study, the variance, the residual and, for a purpose that will appear later, the sum of the logistic transforms for lung cancer and for control groups. The argument of §2.4 shows that the $\hat{\Delta}_j^{(w)}$ can be regarded as estimating also the logistic difference in the proportions not suffering from lung cancer as between two populations, one of non-smokers and one of smokers.

The data provide strong evidence against the constancy of the logistic difference. Some individual residuals are much too large, although that for study 11 is suspect because the number of non-smokers in the lung cancer group is only 4. The residual

TABLE 6.1

*Fourteen retrospective studies on the association between
smoking and lung cancer*

Study	Lung Cancer Patients		Control Patients	
	Total	Non-smokers	Total	Non-smokers
1	86	3	86	14
2	93	3	270	43
3	136	7	100	19
4	82	12	522	125
5	444	32	430	131
6	605	8	780	114
7	93	5	186	12
8	1357	7	1357	61
9	63	3	133	27
10	477	18	615	81
11	728	4	300	54
12	518	19	518	56
13	490	39	2365	636
14	265	5	287	28

TABLE 6.2

Logistic analysis of the data of Table 6.1

Study, j	$\tilde{\Delta}_j^{(w)}$	$v_{\Delta_j}^{(w)}$	Residual	Logistic sum
1	1·83	0·426	0·37	−5·16
2	1·90	0·368	0·53	−5·25
3	1·51	0·214	−0·16	−4·45
4	0·64	0·107	−2·89	−2·96
5	1·74	0·045	0·74	−3·40
6	2·61	0·137	2·76	−6·14
7	0·25	0·298	−2·44	−5·68
8	2·27	0·161	1·71	−8·40
9	1·79	0·391	0·32	−4·55
10	1·37	0·072	−0·79	−5·16
11	3·81	0·274	4·25	−6·85
12	1·18	0·075	−1·50	−5·41
13	1·46	0·030	−0·73	−3·46
14	1·81	0·243	0·46	−6·30
Weighted mean	1·585			

sum of squares is 47·7 to be compared approximately with chi-squared with 13 degrees of freedom.

Although a full interpretation would require detailed knowledge of the individual studies, some of the techniques for a further analysis can be illustrated.

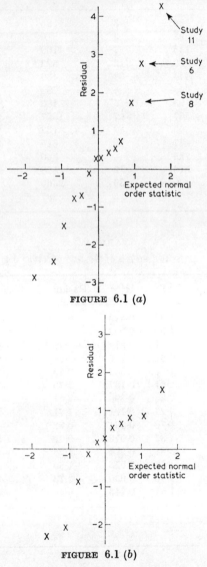

FIGURE 6.1 (*a*)

FIGURE 6.1 (*b*)

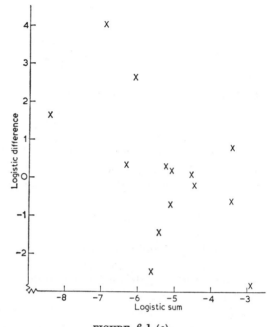

FIGURE 6.1 (c)

FIGURE 6.1 Graphical analyses of data of Table
6.1. (a) Normal plot of residuals (b) Normal plot
of residuals omitting studies 6, 8 and 11. (c) Logistic
difference plotted against logistic sum.

The ranked logistic residuals of Table 6.2 are plotted against
the expected normal order statistics in Fig. 6.1a. Three values,
corresponding to studies 8, 6 and 11, stand out. When these are
omitted, and residuals recalculated from the mean of the remaining eleven studies, Fig. 6.1b is obtained. The resulting residual
sum of squares is 15·1 with 11 degrees of freedom, and, while
there are two rather large negative residuals, much of the
systematic variation has been removed. Unit slope is expected.

For further interpretation the estimated $\varDelta_j^{(w)}$'s can be plotted
against variables characterizing the individual studies. The only
such variable available here is the overall level of response in the
study, conveniently measured by the sum of the logistic transforms in lung cancer and control groups and given in the last
column of Table 6.2. Fig. 6.1c shows the result of plotting the
difference of logistic transforms against the sum. Note that

because the individual transforms have unequal variances, the sums and differences are correlated. Since, however, the variation in the sums is large compared with the individual sampling errors this correlation has no influence on the qualitative interpretation of Fig. 6.1c.

There is some suggestion from Fig. 6.1c that the three studies, 8, 6 and 11, have not only large differences between control and lung cancer groups but also have large negative logistic sums, i.e. overall low proportions of non-smokers. Thus the largest logistic differences between control and lung cancer groups have been found in studies in which the overall proportion of non-smokers is low. Taken with the result of §2.4 this suggests that the largest proportional effect of smoking on lung cancer may be found in contexts in which the overall proportion of smokers is high. Of course, this is no more than a tentative suggestion obtained from an informal graphical analysis and the explanation of the differences between studies may well lie in some other features of the populations under study.

Example 6.2 *A* $2 \times 2 \times 2$ *contingency table.* A particular case of the previous problem arises with the data of Table 6.3 (Woolf, 1955), which can be regarded as a $2 \times 2 \times 2$ table, i.e. as two 2×2 tables identified by the two levels of a third factor. In such situations we have to decide which dimensions of the table represent responses, to be treated as random variables, and which represent factors, to be treated as non-random. In the present example we have what is probably the most common situation. One of the dimensions, blood group A or O, represents a response, which can be regarded as a binary random variable. The other two dimensions define factors. The first factor represents the two types of individual, peptic ulcer patients and controls. The second factor represents the two places where data are available. By the discussion of §2.4, the logistic difference would be the same in a prospective study in which the response is the occurrence of peptic ulcer, this being measured for two sets of individuals of different blood groups.

The logistic differences between peptic ulcer and control sets and the approximate variances of these differences are, by the method of Example 3.1:

London: -0.371, with a variance of 0.00328,
Manchester: -0.200, with a variance of 0.00731.

Thus to test for the constancy of the logistic difference, we calculate

$$\frac{0\cdot171}{\sqrt{0\cdot01058}} \simeq 1\cdot7, \tag{6.12}$$

so that there is some evidence of a difference between London and Manchester.

TABLE 6.3

Blood groups of peptic ulcer and control patients

	London		Manchester	
	Control	Peptic ulcer	Control	Peptic ulcer
Group A	4219	579	3775	246
Group O	4578	911	4532	361
Total	8797	1490	8307	607

If it is assumed that this difference is a sampling deviation, the best single estimate of the logistic effect is a weighted average of the London and Manchester estimates and is thus $-0\cdot318$, with a standard error of $0\cdot048$. Thus the ratio of the odds is estimated as $e^{-0\cdot318} \simeq 0\cdot73$. If in a prospective study the probability of peptic ulcer is small, we thus estimate the ratio of the probability of peptic ulcer for blood group A to that for blood group O to be about $0\cdot73$, for the populations investigated here.
Example 6.3 *Stimulus–response curve.* As an example of a weighted analysis, we consider the fitting of a logistic model to the data of Table 5.3. In the light of the analysis of Example 5.4, we ignore the dependence on u_2 and consider as a single regressor variable the heating time u_1, now denoted by u. We take as numbers of successes and trials the column totals of Table 5.3a.

At the lowest level of u, there are no successes and therefore in the present method that level can be ignored as having zero weight. For this particular type of application in which the different groups are ordered, Berkson (1953) has suggested a special modification of the empirical logistic transform in which the crude transform $\log\{R/(n-R)\}$ is used when $R \neq 0$ or n, but in which $R = 0$ and n are treated as corresponding respectively to $R = \frac{1}{2}$ and $n - \frac{1}{2}$. Here, however, we shall use the generally applicable definition (6.2), the relevant values being shown in Table 6.4. Lower case letters are used to denote observed values of R, Z and V.

TABLE 6.4

Numbers, r_j, of ingots not ready for rolling for various heating times, u_j

u_j	r_j	n_j	$z_j^{(w)}$	$v_j^{(w)}$	$\tilde{\theta}_j^{(w)}$	$\tilde{r}_j^{(w)}$
7	0	55	—	—	0·0079	0·43
14	2	157	−4·635	0·503	0·0134	2·11
27	7	159	−3·149	0·148	0·0355	5·64
51	3	16	−1·609	0·385	0·1870	2·99

The weighted least squares equations (6.4) corresponding to the logistic model $\lambda^{(j)} = \alpha + \beta u_j$ are

$$\begin{pmatrix} 11\cdot3214 & 342\cdot24 \\ 342\cdot24 & 12061\cdot30 \end{pmatrix} \begin{pmatrix} \tilde{\alpha}^{(w)} \\ \tilde{\beta}^{(w)} \end{pmatrix} = \begin{pmatrix} -34\cdot599 \\ -914\cdot487 \end{pmatrix}, \qquad (6.13)$$

leading to

$$\tilde{\alpha}^{(w)} = -5\cdot366, \qquad \tilde{\beta}^{(w)} = 0\cdot0764. \qquad (6.14)$$

For example, in (6.13) the second element on the right-hand side is

$$-4\cdot635 \times 14/0\cdot503 + \cdots.$$

An approximate covariance matrix is the inverse of the matrix on the left-hand side of (6.13), namely

$$\begin{pmatrix} 0\cdot621 & -0\cdot0176 \\ -0\cdot0176 & 0\cdot0005830 \end{pmatrix}. \qquad (6.15)$$

Thus the approximate standard error of $\tilde{\beta}^{(w)}$ is the square root of the second diagonal element of (6.15) and so is 0·0241. The corresponding estimate and standard error using the crude transform $z' = \log\{r/(n-r)\}$ instead of the modified transform $z^{(w)}$ are 0·0748 and 0·0247 respectively.

To check the fit, the simplest procedure is to find the fitted probabilities $\tilde{\theta}_j^{(w)}$ from the fitted transforms $\tilde{\alpha}^{(w)} + \tilde{\beta}^{(w)} u_j$ and thence to calculate fitted numbers of successes, $\tilde{r}_j^{(w)}$. The values given in the final column of Table 6.4 indicate excellent agreement.

6.4 Maximum likelihood; general

We now turn to techniques based on the method of maximum likelihood. Routine application of this method leads to estimates and their asymptotic standard errors and, by comparison of the maximized log likelihoods achieved under different models, to asymptotic chi-squared tests for hypotheses about groups of parameters in the model.

Now, by (2.17), the log likelihood is

$$L(\boldsymbol{\beta}) = \sum_{s=1}^{p} \beta_s T_s - \sum_{i=1}^{n} \log(1 + e^{\mathbf{a}_i\boldsymbol{\beta}}), \qquad (6.16)$$

where

$$T_s = \sum_{i=1}^{n} a_{is} Y_i, \qquad \mathbf{a}_i\boldsymbol{\beta} = \sum_{s=1}^{p} a_{is}\beta_s.$$

Therefore

$$\frac{\partial L(\beta)}{\partial \beta_s} = T_s - \sum_{i=1}^{n} \frac{a_{is} e^{\mathbf{a}_i\boldsymbol{\beta}}}{1 + e^{\mathbf{a}_i\boldsymbol{\beta}}}, \qquad (6.17)$$

$$I_{s_1 s_2}(\boldsymbol{\beta}) \equiv E\left\{-\frac{\partial^2 L(\boldsymbol{\beta})}{\partial \beta_{s_1} \partial \beta_{s_2}}\right\} = \sum_{i=1}^{n} \frac{a_{is_1} a_{is_2} e^{\mathbf{a}_i\boldsymbol{\beta}}}{(1 + e^{\mathbf{a}_i\boldsymbol{\beta}})^2}; \qquad (6.18)$$

note that (6.17) involves the Y_i's only through T_s and that (6.18) does not depend on the Y_i's at all. Thus, in particular, (6.18) gives both the expected and the observed values of the second derivatives of $L(\boldsymbol{\beta})$.

The maximum likelihood estimate $\hat{\boldsymbol{\beta}}$ satisfies the equations

$$\left[\frac{\partial L(\boldsymbol{\beta})}{\partial \beta_s}\right]_{\boldsymbol{\beta}=\hat{\boldsymbol{\beta}}} = 0; \qquad (6.19)$$

and its asymptotic covariance matrix is the inverse matrix $\{I^{s_1 s_2}(\boldsymbol{\beta})\}$ to (6.18) and is consistently estimated by $\{I^{s_1 s_2}(\hat{\boldsymbol{\beta}})\}$. Thus an approximate $(1 - 2\epsilon)$ equitailed confidence interval for, say, β_s is

$$[\hat{\beta}_s - k_\epsilon \sqrt{I^{ss}(\hat{\boldsymbol{\beta}})}, \hat{\beta}_s + k_\epsilon \sqrt{I^{ss}(\hat{\boldsymbol{\beta}})}], \qquad (6.20)$$

where $\Phi(-k_\epsilon) = \epsilon$. To test a hypothesis that the parameter $\boldsymbol{\beta}$ lies in a q-dimensional subspace $\mathscr{B}^*$, for example that only a specified q of the β_s's are non-zero, we calculate

$$L(\hat{\boldsymbol{\beta}}) - L(\hat{\boldsymbol{\beta}}^*; \boldsymbol{\beta} \in \mathscr{B}^*), \qquad (6.21)$$

where the second term is the maximized log likelihood subject to the constraint. If the hypothesis is true, (6.21) is distributed asymptotically as one-half chi-squared with $(p - q)$ degrees of freedom.

This gives an alternative way of finding an approximate confidence region for a particular parameter β_s. For we may take the null hypothesis $\beta_s = \beta_s^{(0)}$ and test it by computing maximized log likelihoods first generally and then subject to the constraint $\beta_s = \beta_s^{(0)}$, i.e. maximizing only with respect to the other $(p - 1)$ components of $\boldsymbol{\beta}$. This leads to the test statistic

$$L(\hat{\boldsymbol{\beta}}) - L(\hat{\boldsymbol{\beta}}^*; \beta_s = \beta_s^{(0)}). \qquad (6.22)$$

We take as the approximate confidence region for β_s the set of all $\beta_s^{(0)}$ not rejected by the test statistic (6.22). That is, if $c_{\nu,\epsilon}^2$ denotes the upper ϵ point of the chi-squared distribution with ν degrees of freedom, the region is

$$\{\beta_s^{(0)}; L(\hat{\boldsymbol{\beta}}) - L(\hat{\boldsymbol{\beta}}^*; \beta_s = \beta_s^{(0)}) \leqslant \tfrac{1}{2} c_{1,\epsilon}^2\}. \qquad (6.23)$$

In a similar way, joint confidence regions for several components can be found.

The regions (6.20) and (6.23) are asymptotically equivalent. The use of (6.23) has the general advantage of being independent of the particular parametrization adopted. For example, if we wrote $\gamma_t = e^{\beta_t}$ $(t = 1, \ldots, p)$ and repeated the arguments leading to (6.20) and (6.23) in terms of the new parameters, the regions based on an asymptotic normal distribution for $\hat{\gamma}_s$ would not correspond exactly to those using a normal distribution for $\hat{\beta}_s$.

To apply any of the above results, the main problem is the computational one of maximizing (6.16); once $\hat{\boldsymbol{\beta}}$ is found the rest is routine. There are numerous procedures for finding numerically the maximum of a relatively complicated function such as (6.16); see, for example, Beale (1967) and Draper and Smith (1966). Usually procedures such as the Newton-Raphson iterative solution of the maximum likelihood equations (6.19) will be reasonably effective, especially if the number of parameters is not too large. If, however, a good general maximization program is available it may well be preferable to use it rather than to develop an *ad hoc* program.

Here we confine the discussion to the following points:

(*i*) how can good first approximations be found from which to start an iterative solution for $\hat{\beta}$;

(*ii*) when are there likely to be numerical difficulties in finding $\hat{\beta}$ and how can such difficulties be avoided?

We shall ignore the possibility that several local maxima give comparable values to the log likelihood; note, however, that in such cases (6.23) would lead to a region consisting of several disjoint intervals. It seems likely on general grounds that multiple maxima will not arise unless there are either very limited data or gross discrepancies with the model.

In some cases an initial estimate can be obtained graphically or by least squares starting with empirical logistic transforms, a somewhat arbitrary grouping of the data being used to calculate the transforms if necessary. Sometimes, however, we can find an approximate solution by manipulation of the maximum likelihood equations themselves.

Suppose first that:

(*i*) all the probabilities of success, $e^{\lambda_i}/(1 + e^{\lambda_i})$, are small;

(*ii*) the variation in the λ_i's is small;

(*iii*) the model contains a constant term.

The argument to be used can best be illustrated by the simple regression model in orthogonalized form, namely

$$\lambda_i = \alpha + \beta(x_i - \bar{x}). \qquad (6.24)$$

Then the log likelihood is

$$\alpha \sum Y_i + \beta \sum (x_i - \bar{x}) Y_i - \sum \log\{1 + e^{\alpha + \beta(x_i - \bar{x})}\}$$
$$\simeq \alpha \sum Y_i + \beta \sum (x_i - \bar{x}) Y_i - e^{\alpha} \sum e^{\beta(x_i - \bar{x})}$$
$$\simeq \alpha \sum Y_i + \beta \sum (x_i - \bar{x}) Y_i - e^{\alpha}\{n + \tfrac{1}{2}\beta^2 \sum (x_i - \bar{x})^2\}.$$

7

The first approximation depends on assumption (i), the second on assumption (ii).

Thus, on differentiating, we have the approximate maximum likelihood equations

$$ne^{\hat{\alpha}} \simeq \sum Y_i, \qquad \hat{\beta}e^{\hat{\alpha}} \sum (x_i - \bar{x})^2 \simeq \sum (x_i - \bar{x}) Y_i. \quad (6.25)$$

Also, to a first order of approximation,

$$\text{var}(\hat{\alpha}) = e^{-\alpha}/n, \quad \text{var}(\hat{\beta}) = e^{-\alpha}\{\sum (x_i - \bar{x})^2\}^{-1}, \quad \text{cov}(\hat{\alpha}, \hat{\beta}) = 0.$$

The close formal connection with the results of ordinary unweighted regression theory is apparent. The same argument in the multiple regression case leads to the analogue of (6.25), involving the matrix of corrected sums of squares and products of the regressor variables.

A second situation when the maximum likelihood equations can be approximated is when none of the fitted probabilities from the model is near zero or one. It is then possible to use the simple approximation

$$\frac{e^t}{1 + e^t} \simeq \begin{cases} 1 & (t > 3), \\ \frac{1}{2} + \frac{1}{6}t & (|t| \leqslant 3), \\ 0 & (t < -3). \end{cases} \quad (6.26)$$

This has a maximum error of 0·07. Therefore, if in the maximum likelihood equations (6.19), $|\mathbf{a}_i \hat{\boldsymbol{\beta}}| < 3$ for all i, we can rewrite the equations as

$$\tfrac{1}{6} \sum a_{is} a_{it} \hat{\beta}_t \simeq \sum (Y_i - \tfrac{1}{2}) a_{is}. \quad (6.27)$$

For the particular linear regression model (6.24), we get

$$\tfrac{1}{6}n\hat{\alpha} \simeq \sum (Y_i - \tfrac{1}{2}), \qquad \tfrac{1}{6} \sum (x_i - \bar{x})^2 \hat{\beta} \simeq \sum (x_i - \bar{x}) Y_i. \quad (6.28)$$

It would be possible to iterate between (6.26) and (6.27), modifying the contribution of any observation for which $|\mathbf{a}_i \hat{\boldsymbol{\beta}}| > 3$. In practice it is more likely that one would use (6.27) merely to obtain a first approximation from which to start a solution of the full equations (6.19).

One general conclusion from both these approximations is that there is likely to be difficulty in finding $\hat{\boldsymbol{\beta}}$ if the columns of $\mathbf{a}$ are nearly linearly dependent. Therefore it may be good to have a preliminary calculation of the formal 'correlation' matrix of the regressor variables, followed if necessary by a linear transformation of the regressor variables to ones more nearly orthogonal in the usual least squares sense.

6.5 Maximum likelihood; application

The main problem arising in applying maximum likelihood methods is, as stated in §6.4, likely to be the computational one of the efficient maximization of functions of several variables. This will not be discussed further here and we merely give two examples to illustrate the main formulae.

Example 6.4 *Analysis of ingot data (continued).* In Example 6.3 we discussed the use of the empirical transform to fit a linear logistic relation for the dependence on heating time of the probability that an ingot is not ready for rolling. Table 6.4 gives the relevant data; in the analysis by maximum likelihood methods direct account can, of course, be taken of the zero observation at the lowest level.

The log likelihood is

$$12\alpha + 370\beta - \{55\log{(1 + e^{\alpha+7\beta})} + 157\log{(1 + e^{\alpha+14\beta})}$$
$$+ 159\log{(1 + e^{\alpha+27\beta})} + 16\log{(1 + e^{\alpha+51\beta})}\}. \qquad (6.29)$$

This is maximized at

$$\hat{\alpha} = -5\cdot415, \qquad \hat{\beta} = 0\cdot0807 \qquad (6.30)$$

and the inverse of the matrix (6.18) of second derivatives at the maximum likelihood point is

$$\begin{pmatrix} 0\cdot5293 & -0\cdot0148 \\ -0\cdot0148 & 0\cdot000500 \end{pmatrix}. \qquad (6.31)$$

These values are to be compared with the values (6.14) and (6.15) obtained from the empirical transforms and, in particular, $\hat{\beta} = 0\cdot0807$ is to be compared with $\hat{\beta}^{(w)} = 0\cdot0764$; the difference is approximately 1/5th of the standard error of estimation.

Part of the discrepancy is accounted for by the omission of the zero observation from the first analysis. If it is left out also from the maximum likelihood analysis, the new estimate is $0\cdot0752$. Little (1968) has shown that the use of crude empirical transforms tends to give a lower slope than the method of maximum likelihood.

In §6.4 two procedures, (6.20) and (6.23), are suggested for obtaining confidence limits. The first and simpler is to use an asymptotic standard error obtained from (6.31). That for $\hat{\beta}$ is $\sqrt{0 \cdot 000500} = 0 \cdot 0224$ and approximate 95% and 98% limits are thus

$$(0 \cdot 037, 0 \cdot 125) \quad \text{and} \quad (0 \cdot 029, 0 \cdot 133). \tag{6.32}$$

To apply (6.23) further calculations are required. The likelihood (6.29) is maximized with respect to α for a range of values of β to give the function

$$L(\hat{\alpha}^*; \beta = \beta^{(0)})$$

shown in Table 6.5.

TABLE 6.5

Ingot data. Log likelihood maximized with respect to α for given β

$\beta^{(0)}$	$L(\hat{\alpha}^*; \beta = \beta^{(0)})$	$\beta^{(0)}$	$L(\hat{\alpha}^*; \beta = \beta^{(0)})$
0·03	−50·12	0·09	−47·77
0·04	−49·28	0·10	−48·06
0·05	−48·61	0·11	−48·55
0·06	−48·11	0·12	−49·22
0·07	−47·80	0·13	−50·08
0·08	−47·69		

At $\beta = \hat{\beta} = 0 \cdot 0807$, $L(\hat{\alpha}^*; \beta = \hat{\beta}) = -47 \cdot 69$.

A confidence region for β is formed, according to (6.23), from those values of $\beta^{(0)}$ giving $L(\hat{\alpha}^*; \beta = \beta^{(0)})$ sufficiently close to the overall maximum. For 95% and 98% regions the allowable differences from the maximum are, from the chi-squared tables with one degree of freedom,

$$\tfrac{1}{2} \times 3 \cdot 841 = 1 \cdot 920 \quad \text{and} \quad \tfrac{1}{2} \times 5 \cdot 412 = 2 \cdot 706.$$

Thus the confidence regions are those giving $L(\hat{\alpha}^*; \beta = \beta^{(0)}) \geqslant 49 \cdot 61$ and $50 \cdot 40$; a graph plotted from Table 6.5 gives the intervals

$$(0 \cdot 036, 0 \cdot 124) \quad \text{and} \quad (0 \cdot 027, 0 \cdot 133). \tag{6.33}$$

As will usually happen, (6.32) and (6.33) are in close agreement.

Both are approximate, but one reason for preferring (6.33) is explained in §6.4.

Example 6.5 *Trend in a* 2×2 *contingency table* (*continued*). In Example 5.5 we began the analysis of the artificial data of Table 5.4, in which two treatments are under comparison and the observations are obtained in serial order. A test of the relevance of such order by the 'exact' methods of Chapter 5 was straightforward, but investigation of the treatment effect in the presence of trend would have required complicated enumeration.

A rather systematic approach entirely by maximum likelihood methods is to fit five linear logistic models in which for the ith observation

$$\text{Model I}: \quad \lambda_i = \mu; \tag{6.34}$$

$$\text{Model II}: \quad \lambda_i = \begin{cases} \mu - \delta & \text{for treatment A,} \\ \mu + \delta & \text{for treatment B}; \end{cases} \tag{6.35}$$

$$\text{Model III}: \lambda_i = \mu + \beta(i - 15); \tag{6.36}$$

$$\text{Model IV}: \lambda_i = \begin{cases} \mu - \delta + \beta(i - 15) & \text{for treatment A,} \\ \mu + \delta + \beta(i - 15) & \text{for treatment B}; \end{cases} \tag{6.37}$$

$$\text{Model V}: \quad \lambda_i = \begin{cases} \mu - \delta + \beta(i - 15) & \text{for treatment A,} \\ \mu + \delta + \gamma(i - 15) & \text{for treatment B}. \end{cases} \tag{6.38}$$

Note that for computational reasons trends with serial number are expressed in terms of $(i - 15)$ rather than in terms of i, and the logistic difference $\varDelta = 2\delta$ is expressed symmetrically in terms of δ (see §6.4). Verbal description of models I to V is unnecessary; the last model is included to cover the possibility of interaction, on a logistic scale, between treatments and serial order, although the possibility of detecting this with the present data is slight.

Table 6.6 summarizes the results of fitting the above models. The adequacy of the different models could be assessed from the significance of additional parameters, but is most neatly seen from the maximized log likelihoods, recalling that when one extra parameter is fitted the significance of the increase in maximized log likelihood is tested as one-half chi-squared with one degree of freedom. For example, in passing from model III to model IV one extra parameter, δ, is introduced and the increase in maximized log likelihood, 0·62, corresponds to a value of chi-squared of 1·24. Alternatively, the square of the ratio of $\hat{\delta}$ to its standard error is 1·25, in close agreement. Other examples of similar close equivalences can be found in the table.

There is no evidence of non-parallelism of the logistic regression lines for A and B, and strong evidence that there is trend with serial order. The estimated logistic difference between B and A, $\varDelta = 2\hat{\delta}$, is slightly greater after adjustment for serial order than before. While well short of statistical significance at an interesting

TABLE 6.6

Fitting of five models by maximum likelihood to the data of Table 5.4

Model	No. of parameters	Maximized log likelihood	Selected estimates and asymptotic standard errors
I	1	−19·71	
II	2	−19·06	$\hat{\delta} = -0\cdot440 \pm 0\cdot39$
III	2	−16·66	$\hat{\beta} = 0\cdot121 \pm 0\cdot053$
IV	3	−16·04	$\hat{\delta} = -0\cdot480 \pm 0\cdot43;$
			$\hat{\beta} = 0\cdot121 \pm 0\cdot055$
V	4	−16·00	$\hat{\delta} = -0\cdot505 \pm 0\cdot45;$
			$\hat{\beta} = 0\cdot139 \pm 0\cdot092;$
			$\hat{\gamma} = 0\cdot111 \pm 0\cdot068$

level, the estimated difference represents quite a large effect, and the confidence limits for $\varDelta$ are wide.

One way of appreciating the meaning of the fitted models is to compute fitted logistic transforms $\hat{\lambda}_i$, and hence fitted probabilities, $\hat{\theta}_i$. The latter are shown for model IV in Table 5.4. Clearly the detailed form of the fitted probabilities depends strongly on the assumptions of the logistic model.

6.6 Examination of goodness of fit

As with other probabilistic models used in the analysis of data, the linear logistic models studied in this monograph are provisional working bases for the analysis rather than rigid specifications to be accepted uncritically. Therefore methods for examining adequacy of fit are important. With relatively large amounts of data graphical analyses and inspection of tables of summary statistics will often lead to a searching examination of the adequacy of a proposed model. With smaller amounts of data, formal tests of significance become relatively more important, primarily because the possibility of apparently appreciable, but nevertheless spurious departures from the model becomes more important.

For analyses based on the empirical logistic transform, methods of examining goodness of fit used in normal-theory problems are applicable with minor modifications. In particular, there are the following possibilities.

(i) The empirical transforms have variances known approximately. Therefore the residual sum of squares can be scaled to correspond, if the model is correct, to unit theoretical variance and hence to be distributed approximately as chi-squared with $(g - p)$ degrees of freedom, where g is the number of groups and p is the number of parameters fitted.

(ii) Residuals can be defined as the differences between the observed and the fitted transforms scaled to have approximately unit variance. These can then be plotted against regressor variables already in the model, against new regressor variables of potential interest and against the fitted values. Further the overall distribution of residuals may indicate the presence of outliers whose inclusion in or exclusion from the main analysis will need consideration. Non-normality of the distribution will, however, be difficult to interpret in the absence of detailed knowledge of the closeness to normality to be expected when the model is correct. Cox and Snell (1968) have given a more complicated definition of residuals that will give a more nearly normal distribution and hence a more nearly linear plot on probability paper. A systematic departure from expectation in any of the above plots will suggest a departure from the initial model; see Anscombe (1961) for a thorough discussion of the properties and applications of residuals in normal-theory situations.

(iii) A more complicated model can be fitted, containing additional parameters representing particular types of departure from the initial model. The statistical significance of the estimates of the additional parameters tests the adequacy of the initial model. In particular, the difference between the residual sums of squares under the two models can be tested approximately as chi-squared with $(p_e - p)$ degrees of freedom, where p_e is the number of parameters fitted in the extended model. As an example, the linearity of a logistic stimulus–response relation $\lambda_i = \alpha + \beta x_i$ can be tested by fitting the relation $\lambda_i = \alpha + \beta x_i + \gamma x_i^2$ and testing the significance of the estimate of γ.

Methods using residuals are of particular value in exploring data where many types of departure from the initial model are of possible importance. Inspection of residuals may then lead to the fitting of an extended model as in (iii), or to a quite new form of model.

If the main analysis is based on maximum likelihood fitting, method (iii) is easily adapted. The simplest analogue of method (i) is obtained by grouping the data in some convenient way, computing observed and fitted numbers of successes for each group and then defining a chi-squared goodness of fit statistic in the usual way by summing over groups

(observed no. of successes − fitted no.)^2/fitted no.

We now consider briefly an analogue of residuals appropriate when no convenient grouping of the data is available. First consider the single binary random variable Y_i having probability of success θ_i. Then the random variable

$$\left(\frac{1 - \theta_i}{\theta_i}\right)^{\frac{1}{2}} Y_i - \left(\frac{\theta_i}{1 - \theta_i}\right)^{\frac{1}{2}} (1 - Y_i) = \frac{Y_i - \theta_i}{\sqrt{\{\theta_i(1 - \theta_i)\}}} \qquad (6.39)$$

has zero mean and unit variance. This is, however, not observable, because θ_i is in general unknown. For a particular model, possibly, but not necessarily, a linear logistic model, θ_i will be a function of unknown parameters $\boldsymbol{\beta}$. Let $\hat{\boldsymbol{\beta}}$ be the maximum likelihood estimate of $\boldsymbol{\beta}$ and write $\hat{\theta}_i = \theta_i(\hat{\boldsymbol{\beta}})$. It then seems sensible to define a residual

$$D_i = \left(\frac{1 - \hat{\theta}_i}{\hat{\theta}_i}\right)^{\frac{1}{2}} Y_i - \left(\frac{\hat{\theta}_i}{1 - \hat{\theta}_i}\right)^{\frac{1}{2}} (1 - Y_i). \qquad (6.40)$$

Where there are, say, m individuals with approximately the same value of $\hat{\theta}_i$ and with approximately the same value of the variables against which plots are to be made, it is sometimes convenient to define a residual for the group as $\sqrt{m}$ times the mean of the separate residuals.

The D_i's can be plotted against original or potential regressor variables. The standardization of the D_i's to have approximately zero mean and unit variance gives some indication of the dispersion to be expected, although the frequency distribution will typically be far from normal. If an approximate test statistic for dependence on a variable z_i is required, it is simple to consider T_z, where

$$T_z = \sum D_i z_i, \qquad (6.41)$$

which will have mean near to zero; its variance can be found from the results of Cox and Snell (1968), allowance being made for the correlation between the D_i's introduced by the use of a common estimate $\hat{\beta}$. With several sets of data, T_z can be computed for each set and a systematic departure from the model will be indicated if the T_z's are predominantly of one sign. Note that T_z is not in general an efficient statistic for testing regression on z_i.

A different procedure is needed to produce the analogue of a plot of residuals against fitted values. A suitable approach is suggested by the analysis of §4.4 for the comparison of a set of binary observations with a corresponding set of probabilities. The distinction in the present section is that the probabilities are estimated from the same data rather than being given constants. With extensive data the best procedure is to sort the individuals into sets with almost constant $\hat{\theta}_i$ and to plot the sample proportion of successes against $\hat{\theta}_i$. A systematic departure from equality indicates a 'non-linearity', e.g. that the small probabilities predicted by the model are systematically too large or too small. The adjustment needed for this may be either to fit a model non-linear in the regressor variables or to change from the logistic function to some other. The two test statistics of §4.4 can be adapted to the present situation, replacing the known p_i's by the fitted $\hat{\theta}_i$'s. In finding the variances of the test statistics it is, however, necessary to take account of errors in the $\hat{\theta}_i$'s (Cox and Snell, 1968).

Example 6.6 *Trend in a* 2×2 *contingency table* (*continued*). The main application of graphical methods using residuals is likely to be in the reduction of extensive data. Here we give a simple example based on the data of Table 5.4, continuing the discussion of Examples 5.5 and 6.5. Residuals can be calculated after the

fitting by maximum likelihood of any of the models of Example
6.5. Here we consider model IV, in which the logistic transform
on the ith trial is

$$\lambda_i = \begin{cases} \mu - \delta + \beta(i - 15) & \text{for treatment A,} \\ \mu + \delta + \beta(i - 15) & \text{for treatment B.} \end{cases}$$

From the maximum likelihood estimates $\hat{\mu}$, $\hat{\beta}$ and $\hat{\delta}$, the fitted
transforms, $\hat{\lambda}_i$, can be calculated, leading to the corresponding
fitted probabilities, $\hat{\theta}_i$. The values, $\hat{\theta}_i$, and the observed residuals
d_i, defined by (6.40), are given in the last two columns of Table
5.4.

The fitted values are of interest in themselves as one clear way
of indicating the meaning of the fitted model. The residuals can
be plotted in various ways, for example against the value of any
other regressor variable that might be relevant. In Fig. 6.2a the
residuals are plotted against serial order, values corresponding
to the two treatments being distinguished.

The absence of trends indicates the adequacy of the model,
although such diagrams have to be interpreted cautiously.
Because of the binary nature of the responses, the residuals,
although scaled to have zero mean and unit standard deviation,
have a very non-normal distribution. In particular, residuals
very close to zero do not occur, except for extreme values of $\hat{\theta}_i$.
In such regions of high or low probability of success, the residuals
have a very skew distribution. Thus if the probability of success
is high, the residuals are either small and positive or large and
negative. The main feature of Fig. 6.2a perhaps calling for
comment is the pair of low values on trials 15 and 17, corresponding to two failures under treatment A in a region of fairly high
probability of success.

The inevitable systematic effects in diagrams such as Fig. 6.2a
can be obviated to some extent by combining residuals in
comparable blocks. In Fig. 6.2b grouped residuals are shown.
They are defined as $\sqrt{3}$ times the average of adjacent sets of
three residuals, for the same treatment. Thus the first three
residuals for treatment A are -0.80, 1.11 and -1.08, and the
grouped residuals give $\sqrt{3} \times (-0.77/3) = -0.44$. The factor $\sqrt{3}$
is included to make the variance approximately one.

We shall not discuss in detail calculation of test statistics
from the residuals. If we wished to test whether the regressions

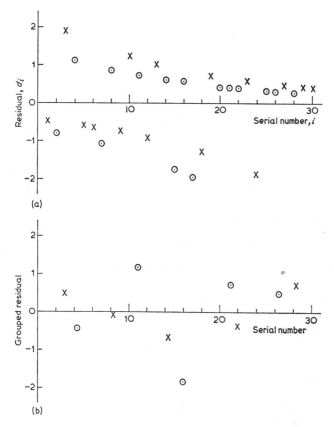

FIGURE 6.2 Two treatments with trend. Plots of residuals.
(a) Individual residuals, d_i: $\odot = A$, $\times = B$. (b) Grouped
residuals: $\odot = A$, $\times = B$.

on serial order have the same logistic slope for the two treatments,
we could calculate a regression coefficient for residuals on serial
order separately for the two treatments, and form a test statistic
from their difference. In the present case this would clearly
confirm the conclusion of Example 6.5, that there is no evidence
from these data that the introduction of separate parameters is
required.

CHAPTER 7

Some further topics

7.1 Introduction

In the previous chapters, methods have been developed for a number of relatively elementary problems connected with binary data. Many extensions of these ideas are possible. Some are outlined in Appendix A while in the present chapter a few others are developed in a little more detail. These are:

(*i*) the relation between binary response probabilities and a continuous distribution of tolerances (§7.2);
(*ii*) the analysis of paired preference data (§7.3);
(*iii*) some problems involving more than two types of response (§§7.4 and 7.5);
(*iv*) some multivariate techniques for binary data (§7.6).

7.2 Tolerances

An alternative way of approaching the simple binary regression problem is occasionally useful. Suppose that a binary response variable Y is observed for a number of pre-assigned values of a regressor variable x, which to be concrete we will call the stimulus. Now suppose that for each individual there is a critical stimulus level, called the tolerance; $Y = 1$ if, and only if, the stimulus applied is above the tolerance. Over a population of individuals, suppose that tolerance is a random variable U, with a probability density function $f(x)$ and a cumulative distribution function $F(x)$. Then if a randomly chosen individual is tested at stimulus x, the probability of success is

$$\text{prob}\,(Y = 1\,;x) = \text{prob}\,(U < x) = F(x). \tag{7.1}$$

Thus if the distribution of tolerances is normal, the binary response curve is of the integrated normal type (2.34). If the binary response function is logistic, then the distribution function of tolerances is

$$\frac{e^{\alpha+\beta x}}{1+e^{\alpha+\beta x}},$$

so that the probability density function is, on differentiating with respect to x,

$$\frac{\beta e^{\alpha+\beta x}}{(1+e^{\alpha+\beta x})^2}.$$

In a reparametized form this is

$$\frac{\exp\{(x-\mu)/\tau\}}{\tau[1+\exp\{(x-\mu)/\tau\}]^2}. \tag{7.2}$$

This is called the logistic probability density function; it is a symmetrical distribution of mean μ and standard deviation $\pi\tau/\sqrt{3}$, having a longer tail than the normal distribution. Statistical analysis of a random sample of continuous observations from (7.2) is not particularly simple, because (7.2) is not within the exponential family of distributions. From a manipulative point of view, the distribution has, however, the property that both density and cumulative distribution functions are expressed explicitly in terms of elementary functions. This is sometimes useful, for example in dealing with censored data.

In particular, one might have a mixture of binary data, observing success or failure at a set of pre-assigned x values, with a second set of data in which a sample of tolerances is measured directly. A combined analysis can be made by maximum likelihood methods, although a graphical or numerical test of the consistency of the two sets of data is desirable.

From (7.1) we can find the binary response curves that correspond to any given tolerance distribution. In some contexts, tolerance might be determined by an event in a Poisson process and then $F(x) = 1 - e^{-\lambda x}$, so that

$$\text{prob}\,(Y=1;x) = 1 - e^{-\lambda x}, \tag{7.3}$$

whence

$$\log \text{prob}\,(Y=0;x) = -\lambda x. \tag{7.4}$$

In other situations, tolerance may have an extreme value distribution,

$$\text{prob}(Y = 1; x) = 1 - \exp\{-e^{-(x-\nu)/\tau}\}, \tag{7.5}$$

implying that

$$\log\{-\log \text{prob}(Y = 0; x)\} = -(x - \nu)/\tau. \tag{7.6}$$

Note that different functions are obtained if successes and failures are interchanged in (7.3) and (7.5). Both (7.3) and (7.5) are sometimes useful in analyzing binary data. Analyses parallel to those using the empirical logistic transform can be based on the linear forms (7.4) and (7.6).

In most situations the concept of a tolerance distribution is unnecessary and it is preferable to work directly with the probabilities of success. The main exceptions occur when the tolerance has an intrinsic physical significance, and when the idea of a tolerance is useful in suggesting models for more complex problems.

7.3 Paired preferences

Suppose that there are m objects or treatments $A_1, \ldots, A_m$ for comparison, and that the observations consist of expressions of preference between pairs of objects; out of n_{ij} comparisons of A_i with A_j, let A_i be preferred r_{ij} times and A_j be preferred $r_{ji} = n_{ij} - r_{ij}$ times. The observations are thus binary, although concerned with responses of different types in different sections of the data.

Denote by π_{ij} the probability that A_i is preferred to A_j. We may hope to be able to express these $\frac{1}{2}m(m-1)$ parameters in terms of a much smaller number; there are a number of approaches to this. With a lot of data we can try to construct a set of m points $P_1, \ldots, P_m$ in a small number d of dimensions such that the relative positions of P_i and P_j determine π_{ij}. With limited data it may be more profitable to assume provisionally that a one-dimensional comparison is involved, i.e. that each object A_i is characterized by a parameter ρ_i and that π_{ij} is a simple function of $\rho_i - \rho_j$. Thus we write

$$\pi_{ij} = g(\rho_i - \rho_j), \qquad \pi_{ji} = 1 - \pi_{ij} = g(\rho_j - \rho_i),$$

so that $g(x) = 1 - g(-x)$ and $g(0) = \frac{1}{2}$. Equivalently the logistic transform is

$$\lambda_{ij} = \log\left(\frac{\pi_{ij}}{\pi_{ji}}\right) = \log\left\{\frac{g(\rho_i - \rho_j)}{1 - g(\rho_i - \rho_j)}\right\} = h(\rho_i - \rho_j), \qquad (7.8)$$

say, where $h(0) = 0$. The simplest form for $h(x)$ is a linear function, $h(x) = ax$, and without loss of generality we may take $a = 1$, this being equivalent to a choice of units on the axis on which the ρ_i's are measured. This leads us to consider

$$\lambda_{ij} = \rho_i - \rho_j, \qquad \pi_{ij} = \frac{e^{\rho_i}}{e^{\rho_i} + e^{\rho_j}} = \frac{\gamma_i}{\gamma_i + \gamma_j}, \qquad (7.9)$$

where $\gamma_i = \log \rho_i$.

The model (7.9) is called the Bradley-Terry model. Many of the statistical problems associated with it have been worked out in detail; an excellent account, dealing also with other methods for handling paired comparison data, is given by David (1963).

The normal-theory analogue of (7.9) concerns an incomplete block design with two plots per block. The within-block analysis of this consists, in effect, of a normal-theory linear model for the differences of the two observations in a block, specifying an expectation $\rho_i - \rho_j$ for a difference between the ith and j^{th} treatments.

The model (7.9) can fail in at least two ways. First, the objects may be arranged one-dimensionally but the function $h(x)$ in (7.8) may be non-linear. This suggests the replacement of (7.9) by

$$\lambda_{ij} = \rho_i - \rho_j + \epsilon(\rho_i - \rho_j)^3, \qquad (7.10)$$

followed by the estimation of ϵ, for example by maximum likelihood methods. The second possibility is that the comparison of the objects is essentially more complex than the comparison of objects arranged in one dimension.

7.4 More than two types of response; ordered responses

In the whole of the previous discussion, binary responses have been considered. We now consider briefly the extent to which the methods can be extended when responses of more than two types are involved.

In some situations there may be three responses, one intermediate. For instance, in a bioassay animals might be classified after test as unaffected, seriously affected and dead. The discussion of tolerance distributions in §7.3 suggests a model that may sometimes be useful. Denote the three responses conventionally by $Y = 0$, 1 and 2. Suppose that the tolerance U and the 'dose' x are such that:

    (*i*) if $U < x$, $Y = 2$ (dead);
    (*ii*) if $U - \Delta/\beta < x \leqslant U$, $Y = 1$ (severely affected);
    (*iii*) if $x \leqslant U - \Delta/\beta$, $Y = 0$ (unaffected).

Then with the logistic density of tolerances, we have that

$$\text{prob}\,(Y = 0) = \frac{1}{1 + e^{\alpha + \Delta + \beta x}}, \qquad \text{prob}\,(Y = 2) = \frac{e^{\alpha + \beta x}}{1 + e^{\alpha + \beta x}},$$

$$\text{prob}\,(Y = 1) = \frac{1}{1 + e^{\alpha + \beta x}} - \frac{1}{1 + e^{\alpha + \Delta + \beta x}}. \tag{7.11}$$

Graphical analyses and tests of (7.11) can be formed in several ways, at least with grouped data. One approach is to consider the binary responses $Y = 0$, $Y \neq 0$ and $Y = 2$, $Y \neq 2$, which should give parallel lines on a logistic scale.

Equation (7.11) refers to a single regressor variable, x, but there is an obvious generalization.

Models analogous to (7.11) are considered for the integrated normal response function by Aitchison and Silvey (1957) and Ashford (1959), who give a maximum likelihood analysis. Gurland, Lee and Dolan (1960) discuss more general forms of tolerance distribution, including the logistic, and give the analysis by weighted least squares.

Special models, such as (7.11) and those of the next section, are to be regarded as suggesting ways in which data may be summarized and analyzed, rather than as rigid specifications to be accepted uncritically.

7.5 More than two types of response; unordered responses

In the previous section, the representation (7.11) treats the possible responses as ordered with $Y = 1$, intermediate between $Y = 0$ and $Y = 2$. We may, however, either have unordered responses, or wish, for some reason, to use a more general model that takes no account of the ordering. One such model is obtained by taking a linear logistic relation for the comparison of the probabilities of any two levels of response. Thus if the possible responses are labelled arbitrarily 0, 1, ..., h, regression on a variable x can be represented by

$$\log \left\{ \frac{\text{prob}\,(Y=a)}{\text{prob}\,(Y=b)} \right\} = (\alpha_a - \alpha_b) + (\beta_a - \beta_b)\,x. \qquad (7.12)$$

This is equivalent to

$$\text{prob}\,(Y=a) = \frac{e^{\alpha_a + \beta_a x}}{\sum\limits_{b=0}^{h} e^{\alpha_b + \beta_b x}} \qquad (a = 0, 1, ..., h). \qquad (7.13)$$

Two constraints can be imposed to make the parameters unique; for example, we may take

$$\alpha_0 = \beta_0 = 0. \qquad (7.14)$$

There is an obvious generalization when there are several regressor variables.

The property of a logistic regression on x is retained in (7.13) when levels of response are pooled only if the parameters β_a are the same in all levels pooled.

A preliminary analysis and test of the model can be based on (7.12). With grouped data a series of linear plots on a logistic scale should be obtained. If one response is relatively frequent it will often be convenient to call it the zero response, to apply (7.14) and to take $\alpha = 0$ in all comparisons (7.12). We thus obtain h lines on a logistic scale.

A generalization of the empirical logistic transform is needed to deal with such problems numerically. Suppose that in one group there are n trials, that the probabilities of the $h + 1$ responses are $\theta_{(0)}, ..., \theta_{(h)}$ and that the corresponding numbers of responses are $R_{(0)}, ..., R_{(h)}$, which are multinomially distributed. Asymptotically

$$E\,\{\log R_{(a)}\} = \log n + \log \theta_{(a)},$$

$$\text{var}\,\{\log R_{(a)}\} = \frac{1 - \theta_{(a)}}{n\theta_{(a)}} \sim \frac{1}{R_{(a)}} - \frac{1}{n}, \qquad (7.15)$$

$$\text{cov}\,\{\log R_{(a)}, \log R_{(b)}\} \sim -\frac{1}{n} \qquad (a \neq b).$$

8

Of the $(h+1)$ frequencies $R_{(a)}$, only h are independent and there is no loss of information in replacing the $\log R_{(a)}$'s by h contrasts which may, for example, be

$$Z_{(a0)} = \log R_{(a)} - \log R_{(0)} \qquad (a = 1, \ldots, h).$$

These have asymptotic expectations

$$\log \{\theta_{(a)}/\theta_{(0)}\} \tag{7.16}$$

and also

$$\text{var} \{Z_{(a0)}\} = \text{var} \{\log R_{(a)}\} - 2\,\text{cov} \{\log R_{(a)}, \log R_{(0)}\}$$
$$+ \text{var} \{\log R_{(0)}\} \sim R_{(a)}^{-1} + R_{(0)}^{-1}, \tag{7.17}$$

$$\text{cov} \{Z_{(a0)}, Z_{(b0)}\} \sim R_{(0)}^{-1} \qquad (a \neq b).$$

Now the generalized models (7.12) and (7.13) specify a linear relation for the asymptotic expectations (7.16) and, the covariance matrix being estimated from (7.17), the method of generalized least squares can be applied for estimation and testing. This approach is set out in detail for the analysis of contingency tables by Plackett (1962) and Goodman (1963a). If $h = 1$, so that the response is binary, the methods reduce to the use of the empirical logistic transform.

More explicitly, a vector of h components $Z_{(0)}$ can be formed from each grouped observation and with k groups of data, a vector Z of hk components is thereby defined. Its asymptotic expectation is, by (7.12) and (7.16), linear in the unknown parameters, say

$$E(Z) \sim a\beta. \tag{7.18}$$

Further the covariance matrix of Z is estimated as

$$V = \text{diag}\,(V_1, \ldots, V_k), \tag{7.19}$$

where V_j is the $h \times h$ matrix obtained by applying (7.17) to the jth group of observations. Thus β is estimated by

$$\tilde{\beta} = (a' V^{-1} a)^{-1} a' V^{-1} Z; \tag{7.20}$$

the covariance matrix of $\tilde{\beta}$ is

$$(a' V^{-1} a)^{-1}. \tag{7.21}$$

The residual sum of squares is

$$(Z - a\tilde{\beta})' V^{-1} (Z - a\tilde{\beta}) \tag{7.22}$$

and gives a test of the adequacy of the model using the chi-squared distribution with $(hk - p)$ degrees of freedom, where p is the number of parameters fitted in (7.18).

If in some groups the frequencies of certain responses are low, it will be necessary to omit the corresponding components from $\mathbf{Z}$, or to use instead the method of maximum likelihood.

7.6 Multivariate responses

We can apply the model of §7.5 when the responses have a factorial structure. The simplest situation of this kind, occurs when there are four levels of response corresponding to a pair (Y_1, Y_2) of binary variables. The four responses can thus be labelled $(0, 0)$; $(0, 1)$; $(1, 0)$; $(1, 1)$. In (7.13) we can write the probabilities as proportional to

$$1; \quad e^{\alpha_{01} + \beta_{01}x}; \quad e^{\alpha_{10} + \beta_{10}x}; \quad e^{\alpha_{11} + \beta_{11}x}. \qquad (7.23)$$

There are corresponding formulae when the factorial structure is more complicated.

One property of (7.23) is that the marginal dependence of Y_1 on x is logistic if and only if $\alpha_{01} + \alpha_{10} = \alpha_{11}$ and $\beta_{01} + \beta_{10} = \beta_{11}$; and this is the condition for the independence of Y_1 and Y_2. When the second condition is satisfied but not the first, there are parallel conditional logistic relations between, say, Y_1 and x given (i) $Y_2 = 0$ and (ii) $Y_2 = 1$.

If (7.23), or a generalization of it, is fitted to data, it will often be helpful to try to describe the α's and the β's in terms of fewer than the full number of parameters. This is broadly analogous to the condensation of the results of factorial experiments in terms of main effects and low order interactions. In fact one way in which condensation can be attempted in the present case is to replace the α's and β's by a smaller number of contrasts. Another possibility is that sets of, say, β's are equal, i.e. that there are groups of responses for which (7.12) is independent of x. In general if there are q binary response variables and p parameters corresponding to regressor variables, we shall in effect obtain in the model p sets of 2^q parameters, only contrasts within any particular set being meaningful in view of (7.14). Although the parameters will be estimated with different precisions, it is legitimate to apply the semi-normal plotting technique (Daniel,

1959) to each of the p sets of estimates as a guide to their condensation.

The objective is to describe the regression of the Y's on the x's as concisely as possible and this can be compared with the aim of canonical correlation and canonical regression analysis in the multivariate analysis of quantitative responses. There the search is for linear combinations of the response variables which have high correlation with the regressor variables. The same idea could formally be applied here, although it is not clear in general that linear combinations of correlated binary variables have particular virtue.

A special case is when the x variable or variables in (7.23) take only two values. In particular with two groups and with two binary responses the probabilities are proportional to

Group 1   1, $e^{\alpha_{01}}$, $e^{\alpha_{10}}$, $e^{\alpha_{11}}$;

Group 2   1, $e^{\alpha_{01}+\Delta_{01}}$, $e^{\alpha_{10}+\Delta_{10}}$, $e^{\alpha_{11}+\Delta_{11}}$.

There is an extensive literature on significance tests in contingency tables in which some dimensions correspond to responses and some to regressor variables; see, for example, the review of Lewis (1962). Such tests are a valuable preliminary to the concise description of data. Another special case is when the linear model represents a linear, or possibly quadratic, surface in some quantitative factor variables. We then have models of response surfaces with multivariate binary responses.

Throughout this discussion, the logistic form is central. Note, however, that a similar family of models can be formed with other functions than exponentials in (7.23) and that given enough data one could choose an appropriate function empirically.

A final generalization concerns joint distributions for mixtures of binary and quantitative responses. One possibility is to take the models of this section as specifying the distribution of the binary components conditionally on the quantitative components, x. If the marginal distribution of the x's is, say, of the multivariate normal form, the combined distribution is of a manageable form for use in problems such as those of discrimination.

Further results and exercises

1. Analyze the data of Table 1.4 by the use of (a) unweighted and (b) weighted empirical logistic transforms. Compare the answers with the results of maximum likelihood fitting as given by Dyke and Patterson (1952) and Cox and Snell (1968).

[§§1.2, 6.2 and 6.4]

2. Obtain estimates of a local linear regression in which the probability of success for the ith trial is $\gamma_0 + \gamma_1 x_i$ and in which ordinary unweighted least squares formulae are applied to the $(0, 1)$ observations. Develop maximum likelihood and least squares formulae for this model (a) not subject to and (b) subject to the constraint that all fitted values are in $[0, 1]$.

[Chapter 2 and §6.4; Tallis, 1964]

3. Examine the following fictitious data comparing two treatments; m is a number so large that sampling fluctuations can be ignored:

	Male		Female	
	Untreated	Treated	Untreated	Treated
Success	$4m$	$8m$	$2m$	$12m$
Total	$7m$	$13m$	$5m$	$27m$

Show that the treatment effect on a logistic scale is the same for males and females, but that if the data for the two sexes are pooled the success rate is the same for treated and untreated individuals. Express algebraically what is involved and comment on the implications for the analysis of binary data.

[§§2.4, 1.2, 5.3; Simpson, 1951; Lindley, 1964]

4. Consider independent observations on a bivariate binary response (Y_1, Y_2), i.e. a 2×2 contingency table with both rows and columns corresponding to random variables, and let $\theta_{ij} =$ prob$(Y_1 = i, Y_2 = j)$, with $\theta_{i.}$ and $\theta_{.j}$ referring to the marginal distributions. Suppose that a measure of association of (Y_1, Y_2) is a function (a) of the pair of conditional probabilities $\theta_{00}/\theta_{0.}$, $\theta_{10}/\theta_{1.}$ and (b) of the pair $\theta_{00}/\theta_{.0}$, $\theta_{01}/\theta_{.1}$. Show that the measure is a function of $(\theta_{00}\theta_{11})/(\theta_{01}\theta_{10})$, i.e. of the logistic difference.

[§2.4; Edwards, 1963]

5. Prove that for the linear logistic model for the 2×2 contingency table, the maximum likelihood estimates and their standard errors are those given by analysis of the crude empirical logistic transforms Z' and their associated variances V' as defined in §3.1. Show that an analogous result holds for any linear logistic model saturated with parameters.

[§§2.4, 3.1 and 6.4]

6. It is required to approximate to the function $e^x/(1 + e^x)$, for $x \geqslant 0$, by (a) a normal distribution function, $\Phi(ax)$ and (b) by a function min$(bx, 1)$. Find the values of the constants a and b giving best fits in the following senses: (i) identity of tangents at $x = 0$; (ii) agreement of standard deviations of full distributions; (iii) Tchebychev (minimax) fit; (iv) least squares fit.

[§2.7]

7. In a binary stimulus–response situation the probability of success at stimulus x is $G(x)$, where $G(x)$ has the mathematical properties of a cumulative distribution function. Prove that by a suitable monotonic transformation of the stimulus scale, say to $z = z(x)$, the probability of success can be taken to be (a) $e^z/(1 + e^z)$, (b) $\Phi(z)$, the standardized normal integral. Show how the transformation for (a) can be estimated from data (i) for arbitrary functions, $G(x)$ and (ii) by assuming that it suffices to take a transformation in the family $z_\rho(x) = (x^\rho - 1)/\rho$ followed by a linear transformation.   [§2.7; Chapter 6]

8. Prove that if R is the number of successes in n independent trials with probability of success θ, then for large n

$$\text{var}\,[\sin^{-1}\{\sqrt{(R/n)}\}] \sim 1/(4n),$$

the angle being measured in radians. Numerical studies show that

$$\sin^{-1}\sqrt{\{R/(n+1)\}} + \sin^{-1}\sqrt{\{(R+1)/(n+1)\}}$$

has a variance within $\pm 6\%$ of $1/(n+\frac{1}{2})$, when $n\theta \geqslant 1$. Examine this analytically by a Taylor expansion about $R = n\theta$. Discuss the circumstances under which a linear model associated with this transformation yields substantial computational advantages. What are its disadvantages, and what are the circumstances under which the more nearly constant variance associated with the second form is a practical advantage?

[§2.7; Freeman and Tukey, 1950]

9. Observations are made on k binary populations classified in a two-way arrangement with k rows and l columns. There are n_{ij} observations in the $(i,j)^{\text{th}}$ cell, the corresponding probability being θ_{ij} with logistic transform λ_{ij}. Obtain sufficient statistics under the additive model $\lambda_{ij} = \mu + \rho_i + \gamma_j$ and show how absence of interactions can be tested. If the n_{ij}'s are equal, or more generally of the form $n_{i.}n_{.j}$, discuss the advantages and limitations of an analysis in terms of an additive model on $\sin^{-1}\sqrt{\theta_{ij}}$. [§2.7; Chapter 5]

10. Obtain, for the situation of Exercise 9, a test broadly analogous to Tukey's degree of freedom for non-additivity, by considering the statistic

$$\sum R_{ij} \frac{R_{i.}\, R_{.j}}{n_{i.}\, n_{.j}},$$

where R_{ij} is the number of successes in the $(i,j)^{\text{th}}$ cell. Show how to obtain, conditionally on $R_{i.} = r_{i.}$, $R_{.j} = r_{.j}$, the mean and variance of the test statistic. [§2.7; Tukey, 1949]

11. In n independent trials with probability of success θ, the observed number of successes is r. If the prior density of θ is proportional to $\theta^a(1-\theta)^b$, show that the posterior density is proportional to $\theta^{a+r}(1-\theta)^{b+n-r}$. Hence show that the posterior density of

$$\left(\frac{n-r+b+1}{r+a+1}\right)\left(\frac{\theta}{1-\theta}\right)$$

has the variance-ratio distribution with $(2r + 2a + 2, 2n - 2r + 2b + 2)$ degrees of freedom. Hence prove that the posterior distribution of $\log\{\theta/(1-\theta)\}$ is approximately normal with mean and variance

$$\log\left(\frac{r+a+\frac{1}{2}}{n-r+b+\frac{1}{2}}\right) \quad \text{and} \quad \frac{1}{r+a+1}+\frac{1}{n-r+b+1}.$$

Show that the methods of analysis using least squares and the empirical logistic transform therefore have a Bayesian interpretation involving independent prior distributions with small a and b.   [Chapter 3; Lindley, 1964]

12. Let R be the number of successes in n independent trials with constant probability of success θ. Prove that

$$E\left\{\psi\left(\frac{R+a}{n+b}\right)\right\} = \psi(\theta) + \frac{a-b\theta}{n}\psi'(\theta) + \frac{\theta(1-\theta)}{2n}\psi''(\theta) + o\left(\frac{1}{n}\right).$$

Examine the special cases $\psi(\theta) = \log\{\theta/(1-\theta)\}$, $\log\theta$ and $\sin^{-1}\sqrt{\theta}$ and hence obtain an approximately unbiased estimate of $\psi(\theta)$. Discuss critically the relevance of the requirement of unbiasedness.   [§3.2]

13. In an experiment at three stimulus levels, there are R_j successes in n_j trials at stimulus level x_j ($j = 1, 2, 3$). Show that if the binary response curve is logistic, the statistic

$$(x_2 - x_3)\log\left(\frac{R_1 + \frac{1}{2}}{n_1 - R_1 + \frac{1}{2}}\right) + (x_3 - x_1)\log\left(\frac{R_2 + \frac{1}{2}}{n_2 - R_2 + \frac{1}{2}}\right)$$
$$+ (x_1 - x_2)\log\left(\frac{R_3 + \frac{1}{2}}{n_3 - R_3 + \frac{1}{2}}\right)$$

has asymptotically zero expectation and a variance estimated by

$$\sum_{i > j \neq k}(x_i - x_j)^2 \frac{(n_k + 1)(n_k + 2)}{n_k(R_k + 1)(n_k - R_k + 1)}.$$

Does this lead to an asymptotically unique test of the adequacy of the logistic model? Give the corresponding test statistic for an integrated normal response curve.

[§3.2; Chambers and Cox, 1967]

14. The independent random variables $Y_1, \ldots, Y_n$ are such that the p.d.f. of Y_i is

$$\exp\{A_i(\theta_i)B_i(y) + C_i(\theta_i) + D_i(y)\},$$

where θ_i is a scalar parameter. A 'linear' model for the θ_i asserts that

$$A_i(\theta_i) = \sum a_{is} \beta_s.$$

Show that there are simple sufficient statistics for the unknown parameters $\boldsymbol{\beta}$, and that much of the discussion of Chapter 4 can be paralleled. Show that for binomial and Poisson distributions the 'linear' models are consistent with natural constraints on the signs of the parameters but that for gamma distributions with known index this is not so.   [Chapter 4; Lehmann, 1959]

15. Individuals are allocated at random to two treatments, a pre-assigned number n_1 receiving treatment 1 and the remaining $n_2 = n - n_1$ receiving treatment 2. On each individual a binary response is then observed. The null hypothesis is that the response observed on any individual is unaffected by the treatment applied to that and other individuals; thus, in particular, the total number r of successes does not depend on the treatment allocation. Show that a test of the null hypothesis based on the randomization distribution of the number of successes in the first treatment is formally identical with Fisher's exact test for the 2×2 contingency table. Show that the other 'exact' tests of Chapter 5 also can be regarded as randomization tests.

[§4.3 and Chapter 5; Barnard, 1947]

16. In the problem of comparing probabilities with observed successes and failures, suppose that a concomitant measurement is available for each individual. How can this be used? Given two sequences $p_1, \ldots, p_n; p_1', \ldots, p_n'$ of probabilities, both of which fit the data adequately, suggest ways of deciding which is the more informative sequence.   [§4.4; Cox, 1958b]

17. In a series of n binary trials the probability of success at any trial, given that there have been v successes in the previous trials is $e^{\alpha+\beta v}/(1 + e^{\alpha+\beta v})$, i.e. each success increases the logistic transform by β. Let j_0 be the serial number of the first success, $j_0 + j_1$ that of the second success, and $j_0 + \ldots + j_{r-1}$ that of the rth and final success; also write $j_r = n - j_0 - \cdots - j_{r-1}$. Prove that the likelihood is

$$\frac{\exp\{r\alpha + \tfrac{1}{2}r(r-1)\beta\}}{\prod\limits_{i=0}^{r} (1 + e^{\alpha+i\beta})^{j_i}}.$$

Hence show that there is no simple sufficient statistic. Show further that if β is large or small, the likelihood involves the j_i's through $T = \sum ij_i$. Prove that when $\beta = 0$ the mean and variance of T are

$$\tfrac{1}{2}(r' - 1)n \quad \text{and} \quad \frac{r'(r' + 1)}{12} \sum (j_i - \bar{j})^2,$$

where $r' = r + 1$ if the series ends with a failure and $r' = r$ otherwise.　　　　　　　　　　　　　　　　　　　[Chapter 5; Cox, 1958a]

18. Find the likelihood for the generalization of Exercise 17 in which each success increases the logistic transform by β and each failure decreases it by γ.　　　　　　　　　　[Chapter 5]

19. Examine an 'exact' procedure for the comparison of two groups adjusting for a concomitant variable z, assuming logistic transforms $\alpha + \beta z$ and $\alpha + \beta z + \Delta$ in the two groups.
　　　　　　　　　　　　　　　　　　　[Chapter 5 and §6.3]

20. Develop a 'random effects' analysis of the paired comparison situation in which for the jth pair, the probabilities of success are the normal integrals with arguments

$$\mu + U_j \quad \text{and} \quad \mu + \Delta + U_j,$$

where U_j is a random variable, characteristic of the jth pair, normally distributed with zero mean and variance σ_u^2, different U_j's being mutually independent. Prove that the probabilities of the four types of response $(0,0); (0,1); (1,0); (1,1)$ are given by integrals of the bivariate normal distribution and hence develop a procedure for estimating Δ, together with a standard error of estimation.　　　　　　　　　　　　　　　　[§5.2]

21. In a matched pair comparison, a concomitant variable u_j is associated with both individuals in the jth pair. Suppose that in the model (5.1) in which a parameter α_j is associated with the jth pair, it is assumed that $\alpha_j = \alpha + \beta u_j$. Discuss the problems of 'exact' and approximate inference about the difference between treatments. How is the analysis modified if the concomitant variable is different for the two individuals in a pair?　　　[§5.2]

22. Set up a linear logistic model for the matched pair problem in which there is an additional parameter representing an effect

of the order in which the two treatments are applied. Obtain a significance test for the treatment effect.   [§5.2; Gart, 1969]

23. Consider a generalization of the matched pair situation in which there are t treatments and the observations are grouped into matched sets of t, one individual in each set receiving each treatment. Set up a logistic model and show that the total numbers of successes per treatment and per set, form the sufficient statistics. An overall test of the equivalence of treatments may be based on the corrected sum of squares of the treatment 'totals'; obtain the mean and variance of the test statistic under the null hypothesis and suggest a chi-squared approximation.   [§5.2; Cochran, 1950]

24. In two 2×2 contingency tables comparing the same two treatments, 1 and 2, the logistic differences between 1 and 2 are Δ_1 and Δ_2. In the jth table the observed numbers of successes are r_{j1} and r_{j2} and the numbers of trials n_{j1} and n_{j2} $(j = 1, 2)$. Show that for inferences about $\Delta = \Delta_2 - \Delta_1$, the distribution for use is

$$p_{R_{22}}(t; \Delta) = \frac{c(\mathbf{n}, \mathbf{r}, t)\, \Delta^t}{\sum_u c(\mathbf{n}, \mathbf{r}, u)\, \Delta^u},$$

where

$$c(\mathbf{n}, \mathbf{r}, t) = \binom{n_{11}}{r_{11} - r_{22} + t}\binom{n_{12}}{r_{11} + r_{12} - t}\binom{n_{21}}{r_{21} + r_{22} - t}\binom{n_{22}}{t}.$$

[§5.3; Bartlett, 1935]

25. Suppose that a large number of 2×2 tables are available comparing the same two treatments. A plot of the estimated logistic difference, Δ, between treatments against the mean logistic transform μ for the two treatments, shows a systematic relation, in which, approximately, $\Delta = g(\mu)$. Show that if the differences Δ are small an analysis of the differences of $h[\log\{\theta/(1 - \theta)\}]$, where $h'(x)g(x) = \text{const}$, leads approximately to constant treatment differences.   [§5.3]

26. In n independent trials the probabilities of success are $\theta_1, \ldots, \theta_n$ and T is the total number of successes. Prove that if and only if $\sum \theta_i(1 - \theta_i)$ diverges as $n \to \infty$, the distribution of T is asymptotically normal. Calculate γ_1 and γ_2 for the distribution of T and show how to use these to improve the normal

approximation. Give an algorithm suitable for computing 'exact' tail areas from the probability generating function of T. [§5.3]

27. Discuss the disadvantages and advantages, if any, of the following methods of analyzing k 2×2 tables, as compared with the methods of §§5.3 and 6.3:

(i) computation of a significance level P_j from the jth table and their combination into the statistic $-2 \sum \log P_j$, to be tested approximately as chi-squared with $2k$ degrees of freedom;

(ii) computation of a standard chi-squared statistic from each table, not corrected for continuity, followed by the testing of the sum as chi-squared with k degrees of freedom;

(iii) computation of a signed chi statistic from each table, followed by a test of their sum as normally distributed, with zero mean and variance k.   [§§5.3 and 6.3; Yates, 1955]

28. Suppose that in an analysis for trend in the probability of success, the null hypothesis is that the probability is a known constant. Formulate this as the problem of testing $\beta = 0$ when the logistic transform of the probability of success on the ith trial is $\alpha_0 + \beta(i - 1)$, and thence obtain the form of the exact test and the mean and variance of the test statistic under the null hypothesis. Compare the test with that appropriate when the initial probability is unknown. Discuss the analysis when there are several sequences, and departures from the null hypothesis may occur both in initial value and in trend, indicating a scheme for graphical and formal analysis.   [§5.4; Cox, 1958a]

29. Prove that for the serial regression problem with n trials and r successes the null probability generating function of the test statistic is

$$\frac{1}{\binom{n}{r}} \prod_{i=1}^{r} \left(\frac{\zeta^i - \zeta^{n+1}}{1 - \zeta^i} \right).$$

[§5.4; Haldane and Smith, 1948]

30. Consider a binary time series in which the probability of success at any trial depends on the outcomes of the previous q trials. Show that for a general q-dependent Markov chain, 2^q parameters have to be estimated, whereas if the logistic analogue

of a qth order autoregressive process is taken, only $q + 1$ parameters are involved. Develop procedures of estimation and testing associated with these two models.   [§5.7]

31. Suppose that in a sequence of trials $\lambda_j = \alpha + \beta' \cos(j\omega) + \beta'' \sin(j\omega)$ for known ω. Show that the sufficient statistics are $\sum Y_j$, $\sum Y_j \cos(j\omega)$ and $\sum Y_j \sin(j\omega)$. If the phase of the relationship is irrelevant, these can be reduced to $\sum Y_j$ and

$$K(\omega) = \{\sum Y_j \cos(j\omega)\}^2 + \{\sum Y_j \sin(j\omega)\}^2,$$

which is proportional to the periodogram ordinate as commonly defined. Conditionally on the observed value of $\sum Y_j$, find the mean and variance of $K(\omega)$ when $\beta' = \beta'' = 0$ and $\omega = 2\pi r/n$, for integral r.   [§5.7]

32. Let R_i be the number of successes in n_i trials, the probability of success $\theta_i(\boldsymbol{\beta})$, depending on the unknown vector parameter $\boldsymbol{\beta}(i = 1, \ldots, k)$. Let $v_i(\theta_i) = n_i \theta_i(1 - \theta_i)$. For a given set $\{\theta_i^*\}$, a least squares estimate of $\boldsymbol{\beta}$, denoted by $\tilde{\boldsymbol{\beta}}(\theta_1^*, \ldots, \theta_k^*)$, is defined by minimizing

$$\sum \frac{\{R_i - n_i \theta_i(\boldsymbol{\beta})\}^2}{v_i(\theta_i^*)},$$

in which minimization the v_i's are not regarded as functions of $\boldsymbol{\beta}$. Show that estimates can be obtained by taking $\theta_i^* = R_i/n_i$ or $\theta_i^* = \theta_i^*(\tilde{\boldsymbol{\beta}}_0)$, where $\tilde{\boldsymbol{\beta}}_0$ is an initial estimate of $\boldsymbol{\beta}$. Prove that the maximum likelihood estimates $\hat{\boldsymbol{\beta}}$ have the latter form with $\hat{\boldsymbol{\beta}} = \tilde{\boldsymbol{\beta}}$, i.e. with weights in conformity with the final estimates.
     [Chapter 6; Berkson, 1955b]

33. Justify the following procedure for fitting a logistic stimulus–response curve to equally spaced stimulus levels. Plot the crude empirical logistic transforms against stimulus level. If the result is not exactly linear, adjust the data as follows: for a suitable stimulus level, subtract $2a$ successes from one stimulus level and add a successes at both neighbouring stimulus levels, where a is a suitable real number, not necessarily an integer. Replot and if necessary repeat the process of adjustment until the plotted points lie effectively on a straight line, the position and slope of the line giving the maximum likelihood estimates.
     [§6.4; Hodges, 1958]

34. Prove that the maximum likelihood equations for the stimulus–response model $\theta_i = e^{\alpha+\beta x_i}/(1 + e^{\alpha+\beta x_i})$ can be written, in the usual notation,

$$\sum R_i = \sum n_i \hat{\theta}_i, \qquad \sum R_i x_i = \sum n_i x_i \hat{\theta}_i,$$

where $\hat{\theta}_i$ is obtained by substituting $\hat{\alpha}$ and $\hat{\beta}$ into the expression for θ_i. Examine the special case where the x_i's are equally and narrowly spaced at intervals h, over a range from very small θ_i to θ_i very near 1, and where all the n_i's are equal. Show, by replacing sums by integrals, that if $\mu = -\alpha/\beta$ is the 50% point, then approximately

$$\hat{\mu} = x_{\max} + \tfrac{1}{2}h - h \sum R_i/n,$$

$$\hat{\mu}^2 + \frac{\pi^2}{3\hat{\beta}^2} = (x_{\max} + \tfrac{1}{2}h)^2 - \tfrac{1}{12}h^2 - 2h \sum R_i x_i/n.$$

Here $\hat{\mu}$ is called the Spearman-Kärber estimate.

[§6.4; Anscombe, 1956]

35. In a stimulus–response situation, there are target stimulus levels $x_1, \ldots, x_n$, there being one observation at each level. The stimuli actually applied are unknown but are independently normally distributed around the targets with variance τ^2. Obtain an expression for the probability of success at target stimulus x, assuming that the logistic regression on actual dose ξ is $\alpha + \beta\xi$. Show that if $\beta\tau$ is small, the logistic transform at target stimulus x is approximately

$$\alpha + \beta x + \tfrac{1}{2}\tau^2 \beta^2 \frac{1 - e^{\alpha+\beta x}}{1 + e^{\alpha+\beta x}}.$$

Discuss the qualitative change in the shape of the response curve. Write down an approximate likelihood and outline an iterative procedure for estimating α and β, regarding τ^2 as known. [§6.4]

36. Show that the situation of the previous exercise is more easily handled when the response relation is of the integrated normal type, when a probability of success at actual level ξ of $\Phi\{(\xi - \mu)/\sigma\}$ is converted into a probability $\Phi\{(x - \mu)/\sigma'\}$ at target level x, where $\sigma'^2 = \sigma^2 + \tau^2$. Prove this result analytically and by an argument based on tolerances. [§§6.4 and 7.2]

37. Contrast the situation of the previous two exercises with that in which there is a random error in measuring the stimulus applied, so that each measured stimulus x_i is normally distributed with variance τ^2 around the corresponding actual stimulus ξ_i, which is unknown. Discuss the difficulties of estimating an assumed logistic regression on ξ, even when τ^2 is known. [§6.4]

38. Suppose that in a binary stimulus–response situation, the probability of success at stimulus x rises monotonically from a lower asymptote ξ, where $\xi \geqslant 0$, to an upper asymptote η, where $\eta \leqslant 1$. Show that a curve of this type is obtained when the probability of success at stimulus x is

$$\frac{\xi + \eta e^{\alpha+\beta x}}{1 + e^{\alpha+\beta x}},$$

where $\beta > 0$. Show how, from suitable data, the parameters ξ, η, α and β can be estimated. How can initial values be obtained for an iterative procedure? What special problems would arise in applying maximum likelihood theory to test the null hypothesis $\xi = 0$? [§6.4; Chernoff, 1954]

39. In n mutually independent binary trials the probability of success is θ_0 for trials $1, \ldots, \nu$ and is θ_1 for trials $\nu + 1, \ldots, n$, where ν is unknown, but θ_0 and θ_1 are known. Find the likelihood and show that the maximum likelihood estimate of ν is obtained by scoring $\log\{(1 - \theta_0)/(1 - \theta_1)\}$ for a failure and $\log(\theta_0/\theta_1)$ for a success. Then $\hat{\nu}$ is such that the cumulative score from observations $1, \ldots, m$, is a maximum at $m = \hat{\nu}$. Set out the corresponding procedure when θ_0 and θ_1 are unknown. [§6.4; Page, 1955]

40. Let R have a binomial distribution corresponding to n trials with probability of success θ. Obtain by expansion the asymptotic skewness of $\phi(Y/n)$, for an arbitrary function $\phi(u)$. Prove that this vanishes if

$$\phi(u) = \int_0^u t^{-\frac{1}{3}}(1 - t)^{-\frac{1}{3}} dt.$$

Hence show that

$$[\phi(Y/n) - \phi\{\theta - \tfrac{1}{6}(1 - 2\theta)/n\}]/\{\theta^{\frac{1}{3}}(1 - \theta)^{\frac{1}{3}}/\sqrt{n}\}$$

may be expected to have a distribution close to the standardized normal distribution. Check this numerically for $n = 5$ and 10; $\theta = 0.1$.   [§6.6; Blom, 1954; Cox and Snell, 1968]

41. Let R_i have a binomial distribution with n_i trials and with probability of success $\theta_i(\boldsymbol{\beta})$, where $\boldsymbol{\beta}$ is an unknown parameter. Let $\hat{\boldsymbol{\beta}}$ be the maximum likelihood estimate of $\boldsymbol{\beta}$ and write $\hat{\theta}_i = \theta_i(\hat{\boldsymbol{\beta}})$. Show that residuals could be defined, for any function $\psi(u)$, by

$$\frac{\psi(R_i/n_i) - \psi(\hat{\theta}_i)}{|\psi'(\hat{\theta}_i)|\{\hat{\theta}_i(1 - \hat{\theta}_i)/n_i\}^{\frac{1}{2}}}.$$

Discuss the relative merits of the choices $\psi(u) = u$, $\psi(u) = \phi(u)$, as given in Exercise 40, and, for linear logistic models, $\psi(u) = \log\{u/(1 - u)\}$.   [§6.6; Cox and Snell, 1968]

42. Show that the continuous logistic density $e^x/(1+e^x)^2$ has moment generating function $s\pi/\sin(s\pi)$ and hence that the odd order cumulants are zero and that the $2r^{\text{th}}$ cumulant is

$$\frac{(-1)^{r+1}\, 2^{2r-1}\, B_{2r}\, \pi^{2r}}{r},$$

where B_{2r} is a Bernoulli number. In particular, show that the standard deviation is $\pi/\sqrt{3}$ and that the coefficient of kurtosis is $6/5$.   [§7.2; Abramowitz and Stegun, 1965, pp. 75, 810]

43. Interpret binary response curves in terms of an underlying continuous distribution in a way alternative to that involving a tolerance distribution. For this, suppose that there is an underlying continuous random variable W and that the binary response $Y = 1$ if and only if $W > 0$. Suppose further that corresponding to stimulus level x, the random variable W has a logistic distribution of mean $\alpha + \beta x$ and unit scale parameter. Generalize this to distributions other than the logistic, and to the case of the general linear model.   [§7.2]

44. Interpret the Bradley-Terry paired comparison model in terms of an underlying continuous logistic distribution of response; assume that the response when P_i is compared with P_j is a logistic variate of unit dispersion and mean $\rho_i - \rho_j$ and that P_i is preferred to P_j if and only if this variate is positive. Show that in an analogous model using normal distributions, the probability that P_i is preferred to P_j is $\Phi(\xi_i - \xi_j)$, called the

Thurstone model. Prove that in the Bradley-Terry model, but not in the Thurstone model, there are simple sufficient statistics for the unknown parameters.   [§7.3; David, 1963, Chapter 4]

45. Show that in the model for three ordered levels of response a consistent model would not be obtained if (7.11) is generalized to allow two different slope parameters in $\text{prob}(Y = 0)$ and $\text{prob}(Y = 2)$.   [§7.4]

46. For three categories of response, compare the models of §§7.4 and 7.5 for regression on a single regressor variable, x.

        [§§7.4 and 7.5]

47. Derive an optimum discriminator between two populations of mixed binary and multivariate normal responses, setting up plausible distributional forms.   [§7.5]

48. Develop a Wald likelihood ratio sequential test for testing the hypothesis that the probability of success is a known constant, against the alternative that there is a linear trend of specified amount on a logistic scale, starting from the known initial value. Obtain also a test for the corresponding problem when the constant probability is an unknown nuisance parameter.   [Wald, 1947; Cox, 1963]

49. Consider two multivariate normal populations, Π_0 and Π_1, in variables $x_1, \ldots, x_p$, the population means being μ_0 and μ_1 and the common covariance matrix Ω. Let the prior probability be π_i that an individual comes from Π_i $(i = 0, 1)$. Show that given the observations $(x_1, \ldots, x_p)$ on one individual, the posterior probability that the individual is from Π_1 satisfies a linear logistic model. Discuss the implications of this for the choice between normal theory discriminant analysis and the analysis of a binary response model.   [Cox, 1966b]

50. Show that results similar to those of Exercise 49 can be obtained for discrimination problems referring to some other members of the exponential family in addition to normal distributions.

Bibliography and references

In the text, references are given only for points stated without proof and to help the reader find further details of topics treated rather sketchily. To have mentioned in the text all papers in any way relevant to the discussion would, I think, have hindered rather than helped the reader. There follows in this Appendix a list of papers referred to in the text and in Appendix A, and a selection of the many other papers dealing with the topics of this monograph.

Abstracts of relevant new work appear in *Mathematical Reviews* and in *Statistical Theory and Methods Abstracts*. For a general account of methods for analyzing qualitative data, with examples mostly from psychological work, see Maxwell (1961).

Work on bioassays and on contingency tables is relevant to the monograph. Both these subjects have very extensive literature, the first being best approached through the books of Finney (1952, 1964) and the second through the review papers of Lewis (1962) and Goodman and Kruskal (1954, 1959).

The importance of the logistic distribution in bioassay and the advantages of estimates obtained by weighted least squares applied to transformed values (minimum logit chi-squared) were stressed in a series of papers by Berkson, (1944, 1951, 1953, 1955a,b, 1957, 1960, 1968).

The logistic function in the context of contingency tables stems from the non-null theory of the 2×2 contingency table (Fisher, 1935) and the analysis of the $2 \times 2 \times 2$ table (Bartlett, 1935). Maximum likelihood fitting of a linear logistic model to a multi-dimensional two-level contingency table was introduced by Dyke and Patterson (1952) and the empirical logistic transform in this context was treated by Woolf (1955).

The account in Chapters 4 and 5 of the systematic use of sufficiency properties to obtain 'exact' procedures is a generalization of Fisher's discussion of the 2×2 contingency table and is a special case of statistical inference for the exponential family (Lehmann, 1959). The present discussion is largely based on the papers by Cox (1958a,b, 1966a,b) and Hitchcock (1966); see also Rasch (1960).

The bibliography that follows consists of books and papers referred to in the text, identified by the relevant section, together with a selected list of other work on topics related to the monograph. Where appropriate, a classification according to the following scheme is used:

Review paper	**Rev**
Adjustment for concomitant variation	**A**
Bayesian methods	**B**
Contingency tables	**C**
Contingency tables, 2×2	**C2^2**
Contingency tables, 2×2 replicated	**C2^{2}r**
Contingency tables, multi-dimensional	**Cm**
Direct analysis of $(0, 1)$ data by least squares	**D**
Errors in response	**Er**
Empirical transforms	**Et**
Matched data	**M**
Multivariate analysis	**Ma**
Maximum likelihood	**Ml**
Multiple responses	**Mr**
Mixtures	**Mx**
Paired preferences	**Pp**
Serial data	**Se**
Sufficiency and 'exact' methods	**Sf**
Transformations, other than logistic	**Tr**
Time series	**Ts**

AFIFI, A. A., and ELASHOFF, R. M. (1969), 'Multivariate two sample tests with dichotomous variables. I. The location model'. *Ann. Math. Statist.* **40**, 290–298.   **Ma**

ABRAMOWITZ, M., and STEGUN, I. A. (1965), *Handbook of Mathematical Functions*, Dover, New York.   [App. A].

AITCHISON, J., and SILVEY, S. D. (1957), 'The generalization of probit analysis to the case of multiple responses', *Biometrika*, **44**, 131–140.   [§7.4].   **Mr**

ANSCOMBE, F. J. (1956), 'On estimating binomial response relations', *Biometrika*, **43**, 461–464.   [§3.2, App. A].   **Et, Ml**

ANSCOMBE, F. J. (1961), 'Examination of residuals', *Proc. 4th Berkeley Symp.*, **1**, 1–36.   [§6.6].

ARMITAGE, P. (1955), 'Tests for linear trends in proportions and frequencies', *Biometrics*, **11**, 375–386.   [§1.2].   **Se**

ARMITAGE, P. (1966), 'The chi-square test for heterogeneity of proportions after adjustment for stratification', *J. R. Statist. Soc.*, **B28**, 150–163.   **A**

ASHFORD, J. R. (1959), 'An approach to the analysis of data for semi-quantal responses in biological response', *Biometrics*, **15**, 573–581.   [§7.4].   **Mr**

ASHFORD, J. R., and SMITH, C. S. (1965a), 'An alternative system for the classification of mathematical models for quantal responses to mixtures of drugs in biological assay', *Biometrics*, **21**, 181–188.   **Mx**

ASHFORD, J. R., and SMITH, C. S. (1965b), 'An analysis of quantal response data in which the measurement of response is subject to error', *Biometrics*, **21**, 811–825.   **Er**

BARNARD, G. A. (1947), 'Significance tests for 2×2 tables', *Biometrika*, **34**, 123–138.   [App. A].   **C2^2**

BARTLETT, M. S. (1935), 'Contingency table interactions', *J. R. Statist. Soc. Suppl.*, **2**, 248–252.   [App. A, B].   **C**

BEALE, E. M. L. (1967), 'Numerical methods', in *Nonlinear Programming*, pp. 135–205. Edited by J. Abadie. N. Holland, Amsterdam.   [§6.4].

BECKER, G. M., DEGROOT, M. H., and MARSCHAK, J. (1963), 'Stochastic models of choice behaviour', *Behav. Sci.*, **8**, 41–55.   **Pp**

BENNETT, B. M. (1956), 'On a rank order test for the equality of the probability of an event', *Skand. Akt.*, **39**, 11–18.   **Se**

BENNETT, B. M. (1964), 'A non-parametric test for randomness in a sequence of multinomial trials', *Biometrics*, **20**, 182–190.   **Se**

BERGER, A. (1961), 'On comparing intensities of association between two binary characteristics in two different populations', *J. Amer. Statist. Assoc.*, **56**, 889–908.   **Cm**

BERKSON, J. (1944), 'Application of the logistic function to bio-assay', *J. Amer. Statist. Assoc.*, **39**, 357–365.   [App. B].   **Et**

BERKSON, J. (1951), 'Why I prefer logits to probits', *Biometrics*, **7**, 327–339.   [App. B].   **Et**

BERKSON, J. (1953), 'A statistically precise and relatively simple method of estimating the bio-assay with quantal response, based on the logistic function', *J. Amer. Statist. Assoc.*, **48**, 565–599.   [§§3.1, 6.3, App. B].   Et

BERKSON, J. (1955a), 'Maximum likelihood and minimum χ^2 estimates of the logistic function', *J. Amer. Statist. Assoc.*, **50**, 130–162.   [§3.5, App. B].   Et, Ml

BERKSON, J. (1955b), 'Estimation by least squares and by maximum likelihood', *Proc. 3rd Berkeley Symp.*, **1**, 1–11. [App. A, B].   Et, Ml

BERKSON, J. (1957), 'Tables for the maximum likelihood estimate of the logistic function', *Biometrics*, **13**, 28–34. [App. B].   Ml

BERKSON, J. (1960), 'Nomograms for fitting the logistic function by maximum likelihood', *Biometrika*, **47**, 121–141.   [App. B].   Ml

BERKSON, J. (1968), 'Application of minimum logit χ^2 to a problem of Grizzle with a notation on the problem of no interaction', *Biometrics*, **24**, 75–95.   [App. B].   Et

BHAPKAR, V. P. (1961), 'Some tests for categorical data', *Ann. Math. Statist.*, **32**, 72–83.   C

BHAPKAR, V. P. (1968), 'On the analysis of contingency tables with a quantitative response', *Biometrics*, **24**, 329–338.   C

BHAPKAR, V. P., and KOCH, G. G. (1968), 'On the hypotheses of no interaction in contingency tables', *Biometrics*, **24**, 567–594.   C

BHAT, B. R., and KULKARNI, S. R. (1966), 'Lamp tests of linear and loglinear hypotheses in multinomial experiments', *J. Amer. Statist. Assoc.*, **61**, 236–245.   C

BILLEWICZ, W. Z. (1956), 'Matched pairs in sequential trials for significance of a difference between proportions', *Biometrics*, **12**, 283–300.   M

BILLINGSLEY, P. (1961), 'Statistical methods in Markov chains', *Ann. Math. Statist.*, **32**, 12–40.   [§5.7].   Rev, Ts

BIRCH, M. W. (1963), 'Maximum likelihood in three-way contingency tables', *J. R. Statist. Soc.*, **B25**, 220–233.   Cm, Ml

BIRCH, M. W. (1964), 'The detection of partial association, I: the 2 × 2 case', *J. R. Statist. Soc.*, **B26**, 313–324.   C2², Sf

BIRCH, M. W. (1965), 'The detection of partial association, II: the general case', *J. R. Statist. Soc.*, **B27**, 111–124.   C, Sf

BLOCH, D. A., and WATSON, G. S. (1967), 'A Bayesian study of the multinomial distribution', *Ann. Math. Statist.*, **38**, 1423–1435.   B

BLOM, G. (1954), 'Transformations of the binomial, negative binomial, Poisson and χ^2 distributions', *Biometrika*, **41**, 302–316.   [App. A].   Tr

BROSS, I. (1954), 'Misclassification in 2×2 tables', *Biometrics*, **10**, 478–486.   C2², Er

BROSS, I. (1964), 'Taking a covariable into account', *J. Amer. Statist. Assoc.*, **59**, 725–736.   C2², A

BRUNK, H. D. (1955), 'Maximum likelihood estimates of monotone parameters', *Ann. Math. Statist.*, **26**, 607–616.   Ml

CHAMBERS, E. A., and COX, D. R. (1967), 'Discrimination between alternative binary response models', *Biometrika*, **54**, 573–578.   [App. A].   Tr

CHAPMAN, D. G., and NAM, J. (1968), 'Asymptotic power of chi square tests for linear trends in proportions', *Biometrics*, **24**, 315–328.   Se

CHASE, G. R. (1968), 'On the efficiency of matched pairs in Bernouilli trials', *Biometrika*, **55**, 365–369.   M

CHERNOFF, H. (1954), 'On the distribution of the likelihood ratio', *Ann. Math. Statist.*, **25**, 573–578.   [App. A].

CLARINGBOLD, P. J., BIGGERS, J. D., and EMMENS, C. W. (1953), 'The angular transformation in quantal analysis', *Biometrics*, **9**, 467–484.   [§2.7].   Tr

COCHRAN, W. G. (1950), 'The comparison of percentages in matched samples', *Biometrika*, **37**, 256–266.   [App. A].   M

COCHRAN, W. G. (1954), 'Some methods for strengthening the common X^2 tests', *Biometrics*, **10**, 417–451.   C

COCHRAN, W. G. (1968), 'The effectiveness of adjustment by subclassification in removing bias in observational studies', *Biometrics*, **24**, 295–313.   A

CORNFIELD, J. (1956), 'A statistical problem arising from retrospective studies', *Proc. 3rd Berkeley Symp.*, **4**, 135–148.   [§§1.2, 6.3].   C2²r

CORNFIELD, J., and HAENSZEL, W., (1960) 'Some aspects of retrospective studies', *J. Chronic Diseases*, **11**, 523–534.   C2²r

COX, D. R. (1958a), 'The regression analysis of binary sequences (with discussion)', *J. R. Statist. Soc.*, **B20**, 215–242.   [App. A, B].   Sf, C2², Se

COX, D. R. (1958b), 'Two further applications of a model for binary regression', *Biometrika*, **45**, 562–565.   [App. A, B].   Sf, M

COX, D. R. (1963), 'Large sample sequential tests of composite hypotheses', *Sankhyā*, **A25**, 5–12.   [App. A].

COX, D. R. (1966a), 'A simple example of a comparison involving quantal data', *Biometrika*, **53**, 215–220. [§5.3, App. B].   C2²r

COX, D. R. (1966b), 'Some procedures connected with the logistic qualitative response curve', *Research Papers in Statistics: Essays in Honour of J. Neyman's 70th Birthday*, pp. 55–71. Edited by F. N. David. Wiley, London. [App. A, B].   Rev, Sf, Ml, Ma

COX, D. R., and LAUH, E. (1967), 'A note on the graphical analysis of multi-dimensional contingency tables', *Technometrics*, **9**, 481–488.   [§3.4].   Cm

COX, D. R., and LEWIS, P. A. W. (1966), *The Statistical Analysis of Series of Events*, Methuen, London.   [§1.2].

COX, D. R., and SNELL, E. J. (1968), 'A general definition of residuals (with discussion)', *J. R. Statist. Soc.*, **B30**, 248–275.   [§6.6, App. A].

CRAMER, E. M. (1964), 'Some comparisons of methods of fitting the dosage response curve for small samples', *J. Amer. Statist. Assoc.*, **59**, 779–793.   Ml, Et

DANIEL, C. (1959), 'Use of half-normal plots in interpreting factorial two-level exponents', *Technometrics*, **1**, 311–341.   [§§3.4, 7.6].

DARROCH, J. N. (1962), 'Interactions in multi-factor contingency tables', *J. R. Statist. Soc.*, **B24**, 251–263.   Cm

DAVID, H. A. (1963), *The Method of Paired Comparisons*, Griffin, London.   [§7.3, App. A].   Rev, Pp

DORN, H. F. (1954), 'The relationship of cancer of the lung and the use of tobacco', *Amer. Statistician*, **8**, 7–13. [§6.3].

DRAPER, N. R., and SMITH, H. (1966), *Applied Regression Analysis*. Wiley, New York.   [§6.4].

DURBIN, J. (1960), 'Estimation of parameters in time-series regression models', *J. R. Statist. Soc.*, **B22**, 139–153. [§3.5].

DYKE, G. V., and PATTERSON, H. D. (1952), 'Analysis of factorial arrangements when the data are proportions', *Biometrics*, **8**, 1–12.   [§§1.2, 3.4, App. A, B].   Ml, Cm

EDWARDS, A. W. F. (1963), 'The measure of association in a 2×2 table', *J. R. Statist. Soc.*, **A126**, 109–114. [App. A].   C2²

ELASHOFF, J. D., ELASHOFF, R. M., and GOLDMAN, G. E. (1967), 'On the choice of variables in classification problems with dichotomous variables', *Biometrika*, **54**, 668–670.   Ma

FELDSTEIN, M. S. (1966), 'A binary variable multiple regression method of analysing factors affecting perinatal mortality and other outcomes of pregnancy', *J. R. Statist. Soc.*, **A129**, 61–73. [§1.2].   **D**

FELLER, W. (1968), *An Introduction to Probability Theory and its Applications*, vol. 1, 3rd ed., Wiley, New York. [§4.3].

FINNEY, D. J. (1952), *Probit Analysis*, 2nd ed., Cambridge University Press. [§2.7, App. B].

FINNEY, D. J. (1964). *Statistical Method in Biological Assay*, 2nd ed., Griffin, London. [App. B].

FISHER, R. A. (1935), 'The logic of inductive inference (with discussion)', *J. R. Statist. Soc.*, **98**, 39–54. Reprinted, without discussion, in Fisher, R. A. (1950). *Contributions to Mathematical Statistics*. Wiley, New York. [App. B]. $C2^2$

FISHER, R. A. (1954), 'The analysis of variance with various binomial transformations (with discussion)', *Biometrics*, **10**, 130–139.   **Ml, Tr**

FISHER, R. A. (1956), *Statistical Methods and Scientific Inference*. Oliver and Boyd, Edinburgh. [§4.2].

FIX, E., and HODGES, J. L. (1955), 'Significance probabilities of the Wilcoxon test', *Ann. Math. Statist.*, **26**, 301–312. [§5.4].

FREEMAN, M. F., and TUKEY, J. W. (1950), 'Transformations related to the angular and the square root', *Ann. Math. Statist.*, **21**, 607–611. [App. A].   **Tr**

GABRIEL, K. R. (1963), 'Analysis of variance of proportions with unequal frequencies', *J. Amer. Statist. Assoc.*, **58**, 1133–1157.   **D**

GART, J. J. (1962a), 'Approximate confidence levels for the relative risk', *J. R. Statist. Soc.*, **B24**, 454–463. [§4.3]. $C2^2$

GART, J. J. (1962b), 'On the combination of relative risks', *Biometrics*, **18**, 601–610. $C2^2r$

GART, J. J. (1966), 'Alternative analyses of contingency tables', *J. R. Statist. Soc.*, **B28**, 164–179.   **C**

GART, J. J. (1969), 'An exact test for comparing matched proportions in crossover designs', *Biometrika*, **56**, [App. A].   **M**

GART, J. J., and ZWEIFEL, J. R. (1967), 'On the bias of various estimators of the logit and its variance with application to quantal bioassay', *Biometrika*, **54**, 181–187. [§3.2].   **Et**

GOOD, I. J. (1963), 'Maximum entropy for hypothesis formulation especially for multidimensional contingency tables', *Ann. Math. Statist.*, **34**, 911–934.   C

GOOD, I. J. (1965), *The Estimation of Probabilities*, Mass. Inst. of Tech. Press, Cambridge, Mass.   B

GOOD, I. J. (1967). 'A Bayesian significance test for multinomial distributions (with discussion)', *J. R. Statist. Soc.*, **B29**, 399–431.   B

GOODMAN, L. A. (1963a), 'On Plackett's test for contingency table interactions', *J. R. Statist. Soc.*, **B25**, 179–188. [§7.5].   C, Et

GOODMAN, L. A. (1963b), 'On methods for comparing contingency tables', *J. R. Statist. Soc.*, **A126**, 94–108.   C

GOODMAN, L. A. (1964), 'Simple methods of analyzing three-factor interaction in contingency tables', *J. Amer. Statist. Assoc.*, **59**, 319–352.   Cm

GOODMAN, L. A., and KRUSKAL, W. H. (1954), 'Measures of association for cross classifications', *J. Amer. Statist. Assoc.*, **49**, 732–764.   [App. B].   Rev, C

GOODMAN, L. A., and KRUSKAL, W. H. (1959), 'Measures of association for cross classifications. II. Further discussion and references', *J. Amer. Statist. Assoc.*, **54**, 123–163.   [App. B].   Rev, C

GOODMAN, L. A., and KRUSKAL, W. H. (1963), 'Measures of association for cross classifications. III. Approximate sampling theory', *J. Amer. Statist. Assoc.*, **58**, 310–364.   C

GORDON, T., and FOSS, B. M. (1966), 'The role of stimulation in the delay of onset of crying in the new-born infant', *J. Exp. Psychol.*, **16**, 79–81.   [§1.2].

GRIZZLE, J. E. (1961). 'A new method of testing hypotheses and estimating parameters for the logistic model', *Biometrics*, **17**, 372–385.   Et

GRIZZLE, J. E. (1962), 'Asymptotic power of tests of linear hypotheses using the probit and logit transformations', *J. Amer. Statist. Assoc.*, **57**, 877–894.   Et

GURLAND, J., LEE, I., and DOLAN, P. A. (1960), 'Polychotomous quantal response in biological assay', *Biometrics*, **16**, 382–398.   [§7.4].   Mr

HALDANE, J. B. S. (1955), 'The estimation and significance of the logarithm of a ratio of frequencies', *Ann. Hum. Genetics*, **20**, 309–311.   [§3.2].   Et

10

HALDANE, J. B. S., and SMITH, C. A. B. (1948), 'A simple exact test for birth-order effect', *Ann. Eugenics*, **14**, 117–124.   [App. A].   Se

HANNAN, J., and HARKNESS, W. (1963), 'Normal approximation to the distribution of two independent binomials, conditioned on fixed sum', *Ann. Math. Statist.*, **34**, 1593–1595.   C2²

HARKNESS, W., and KATZ, L. (1964), 'Comparison of power functions for the test of independence in 2×2 contingency tables', *Ann. Math. Statist.*, **35**, 1115–1127.   C2²

HEWLETT, P. S., and PLACKETT, R. L. (1964), 'A unified theory for quantal responses to mixtures of drugs: competitive action', *Biometrics*, **20**, 566–575.   Mx

HINZ, P., and GURLAND, J. (1968), 'A method of analyzing untransformed data from the negative binomial and other contagious distributions', *Biometrika*, **55**, 163–170. Tr

HITCHCOCK, S. E. (1962), 'A note on the estimation of the parameters of the logistic function, using the minimum logit χ^2 method', *Biometrika*, **49**, 250–252.   Et

HITCHCOCK, S. E. (1966), 'Tests of hypotheses about the parameters of the logistic distribution', *Biometrika*, **53**, 535–544.   [§5.5, App. B].   Sf

HODGES, J. L. (1958), 'Fitting the logistic by maximum likelihood', *Biometrics*, **14**, 453–461.   [App. A].   Ml

KASTENBAUM, M. A., and LAMPHIEAR, D. E. (1959), 'Calculation of chi-square to test the no three-factor interaction hypothesis', *Biometrics*, **15**, 107–115.   Cm

KENDALL, M. G., and STUART, A. (1963), *The Advanced Theory of Statistics*, vol. 1, 2nd ed., Griffin, London. [§5.4].

KENDALL, M. G., and STUART, A. (1967), *The Advanced Theory of Statistics*, vol. 2, 2nd ed., Griffin, London. [§5.4].

KINCAID, W. M. (1962), 'The combination of $2 \times m$ contingency tables', *Biometrics*, **18**, 224–228.   Cr

KU, H. H., and KULLBACK, S. (1968), 'Interaction in multi-dimensional contingency tables: an information theoretic approach', *J. Res. Nat. Bur. St.*, **72B**, 159–199.   Cm

KULLBACK, S., KUPPERMAN, M., and KU, H. H. (1962), 'Tests for contingency tables and Markov chains', *Technometrics*, **4**, 573–608.   Rev, C, Ts

LANCASTER, H. O. (1951), 'Complex contingency tables treated by partition of χ^2', *J. R. Statist. Soc.*, **B13**, 242–249.   Cm

LEHMANN, E. L. (1959), *Testing Statistical Hypotheses*, Wiley, New York.   [§§4.1, 4.2, App. A, B].

LEWIS, B. N. (1962), 'On the analysis of interaction in multidimensional contingency tables', *J. R. Statist. Soc.*, **A125**, 88–117.   [§7.6, App. B].   Rev, Cm

LINDLEY, D. V. (1964), 'The Bayesian analysis of contingency tables', *Ann. Math. Statist.*, **35**, 1622–1643. [App. A].   B, C

LITTLE, R. E. (1968), 'A note on estimation for quantal response data', *Biometrika*, **55**, 578–579.   [§6.5].   Et, Ml

LOMBARD, H. L., and DOERING, C. R. (1947), 'Treatment of the four-fold table by partial correlation as it relates to Public Health problems', *Biometrics*, **3**, 123–128. [§1.2].

MACNAUGHTON-SMITH, P. (1965), *Some Statistical and Other Numerical Techniques for Classifying Individuals*. Studies in the causes of delinquency and the treatment of offenders, HMSO, London.   Ma

MCNEMAR, Q. (1947), 'Note on the sampling error of the differences between correlated proportions or percentages', *Psychometrika*, **12**, 153–157.   M

MANTEL, N. (1963), 'Chi-square tests with one degree of freedom: extensions of the Mantel-Haenzel procedure', *J. Amer. Statist. Assoc.*, **58**, 690–700.   Cr

MANTEL, N. (1966), 'Models for complex contingency tables and polychotomous dosage response curves', *Biometrics*, **22**, 83–95.   Sf, Mr

MANTEL, N., and HAENSZEL, W. (1959), Statistical aspects of the analysis of data from retrospective studies of disease', *J. Nat. Cancer Inst.*, **22**, 719–748.   Cr

MANTEL, N., and HALPERIN, M. (1963), 'Analyses of birth-rank data', *Biometrics*, **19**, 324–340.   Se

MAXWELL, A. E. (1961), *Analysing Qualitative Data*. Methuen, London.   [§5.2, App. B].   Rev

MIELKE, P. W., and SIDDIQUI, M. M. (1965), 'A combinatorial test for independence of dichotomous responses', *J. Amer. Statist. Assoc.*, **60**, 437–441.   Ma

MIETTINEN, O. S. (1968), 'The matched pairs design in the case of all-or-none responses', *Biometrics*, **24**, 339–352.   M

MOSTELLER, F. (1968), 'Association and estimation in contingency tables', *J. Amer. Statist. Assoc.*, **63**, 1–28.   Rev, C

NAYLOR, A. F. (1964), 'Comparisons of regression constants fitted by maximum likelihood to four common transformations of binomial data', *Ann. Hum. Genet.*, **27**, 241–246.   [§2.7].   Tr, Ml

NEWELL, D. J. (1963), 'Misclassification in 2×2 tables', *Biometrics*, **19**, 187–188.   C2², Er

NEYMAN, J. (1949), 'Contribution to the theory of the χ^2 test', *Proc. Berk. Symp.*, 239–273.   Et

OLKIN, I., and TATE, R. F. (1961), 'Multivariate correlation models with mixed discrete and continuous variables', *Ann. Math. Statist.*, **32**, 448–465.   Ma

PAGE, E. S. (1955), 'A test for a change in a parameter occurring at an unknown point', *Biometrika*, **42**, 523–527.   [App. A].

PATNAIK, P. B. (1948), 'The power function of the test for the difference between two proportions in a 2×2 table', *Biometrika*, **35**, 157–175.   C2²

PEARSON, E. S. (1947), 'The choice of statistical tests illustrated on the interpretation of data classed in a 2×2 table', *Biometrika*, **34**, 139–167.   [§4.3].   C2²

PEARSON, E. S., and HARTLEY, H. O. (1966), *Biometrika Tables for Statisticians*, 3rd ed., Cambridge University Press.   [§§4.3, 5.4, 6.3].

PLACKETT, R. L. (1960), *Principles of Regression Analysis*, Clarendon Press, Oxford.   [§2.1].

PLACKETT, R. L. (1962), 'A note on interactions in contingency tables', *J. R. Statist. Soc.*, **B24**, 162–166. [§7.5].   C, Et

PLACKETT, R. L., and HEWLETT, P. S. (1967), 'A comparison of two approaches to the construction of models for quantal responses to mixtures of drugs', *Biometrics*, **23**, 27–44.   Mx

RAO, C. R. (1965), *Linear Statistical Inference and Its Applications*, Wiley, New York.   [§2.1].

RAO, P. V., and KUPPER, L. L. (1967), 'Ties in paired-comparison experiments: a generalization of the Bradley-Terry model', *J. Amer. Statist. Assoc.*, **62**, 194–204.   Pp, Mr

RASCH, G. (1960), *Probabilistic Models for Some Intelligence and Attainment Tests*, Nielson and Lydiche, Copenhagen. [App. B].   Sf

REIERSØL, O. (1961), 'Linear and non-linear multiple comparisons in logit analysis', *Biometrika*, **48**, 359–365.   Et

RIES, P. N., and SMITH, H. (1963), 'The use of chi-square for preference testing in multidimensional problems', *Chemical Engineering Progress*, **59**, 39–43.   [§3.4].   **Cm, Pp**

ROY, S. N., and BHAPKAR, V. P. (1960), 'Some non-parametric analogs of "normal" anova, manova, and of studies in "normal" association', *Contributions to Probability and Statistics*, pp. 371–387. Edited by I. Olkin *et al.* Stanford University Press.   **C**

ROY, S. N., and KASTENBAUM, M. A. (1965), 'On the hypothesis of "no interaction" in a multiway contingency table', *Ann. Math. Statist.*, **27**, 749–757.   **Cm**

ROY, S. N., and MITRA, S. K. (1956), 'An introduction to some non-parametric generalizations of analysis of variance and multivariate analysis', *Biometrika*, **43**, 361–376.   **Cm**

SCHEFFÉ, H. (1959), *The Analysis of Variance*, Wiley, New York.   [§2.1].

SILVERSTONE, H. (1957), 'Estimating the logistic curve', *J. Amer. Statist. Assoc.*, **52**, 567–577.   **Sf**

SIMPSON, E. H. (1951), 'The interpretation of interaction in contingency tables', *J. R. Statist. Soc.*, **B13**, 238–241. [App. A].   **Cm**

SNELL, E. J. (1964), 'A scaling procedure for ordered categorical data', *Biometrics*, **20**, 592–607.   **Ma**

STERLING, T., GLESER, M., HABERMAN, S., and POLLACK, S. (1966), 'Robot data screening: a solution to multivariate type problems in the biological and social sciences', *Communications of A. C. M.*, **9**, 529–532.   **Cm**

STEVENS, W. L. (1939), 'Distribution of groups in a sequence of alternatives', *Ann. Eugenics*, **9**, 10–17.   [§5.7].

STUART, A. (1953), 'The estimation and comparison of strengths of association in contingency tables', *Biometrika*, **39**, 105–110.   **C**

STUART, A. (1957), 'The comparison of frequencies in matched samples', *Brit. J. Statist. Psychol.*, **10**, 29–32.   **M**

TALLIS, G. M. (1964), 'The use of models in the analysis of some classes of contingency tables', *Biometrics*, **20**, 832–839.   [App. A].   **Tr**

TAYLOR, W. F. (1953), 'Distance functions and regular best asymptotically normal estimates', *Ann. Math. Statist.*, **24**, 85–92.   **Et**

TUKEY, J. W. (1949), 'One degree of freedom for non-additivity', *Biometrics*, **5**, 232–242.   [App. A].

WALD, A. (1947), *Sequential Analysis*, Wiley, New York. [App. A].

WALD, A., and WOLFOWITZ, J. (1940), 'On a test of whether two samples are from the same population', *Ann. Math. Statist.*, **11**, 147–162. Reprinted in Wald, A. (1955), *Selected Papers in Statistics and Probability*, McGraw Hill, New York. [§5.7].

WALKER, S. H., and DUNCAN, D. B. (1967), 'Estimation of the probability of an event as a function of several independent variables', *Biometrika*, **54**, 167–179. [§1.2].   Ml

WALSH, J. E. (1963), 'Loss in test efficiency due to misclassification for 2 × 2 tables', *Biometrics*, **19**, 158–162.   C2², Er

WHITTLE, P. (1955), 'Some distribution and moment formulae for the Markov chain', *J. R. Statist. Soc.*, **B17**, 235–242. [§5.7].   Ts

WOOLF, B. (1955), 'On estimating the relation between blood group and disease', *Ann. Hum. Genetics*, **19**, 251–253. [§6.3, App. B].   Et

WORCESTER, J. (1964), 'Matched samples in epidemological studies', *Biometrics*, **20**, 840–848.   M

YATES, F. (1955), 'The use of transformation and maximum likelihood in the analysis of quantal experiments involving two treatments', *Biometrika*, **42**, 382–403. [App. A].   Ml

YATES, F. (1961), 'Marginal percentages in multiway tables of quantal data with disproportionate frequencies', *Biometrics*, **17**, 1–9.

Author Index

Subject Index

142 SUBJECT INDEX

Semi normal plot
  see Half normal plot
Sequential tests, 121
Serial order, 6, 7, 30, 61, 64, 116
Smoking and lung cancer, 2, 3, 22, 23, 35, 80–84
Spearman-Kärber estimate, 118
Statistic, sufficient
  see Sufficient statistic
Stimulus-response curve, 8–10, 25, 30, 32, 42, 85–87
  logistic, definition of, 25
Subjective probability
  agreement with objective, 54
Sufficient statistic, 2, 19, 29, 44, 45, 47, 52, 53, 80, 123
  not complete, 73, 74
Survey
  prospective, 22
  retrospective, 2, 22, 51, 80–84

Threshold, 18
Thurstone model, 121
Time series, 12, 13, 16, 72–75, 113, 114, 116, 117
Tolerances, 100–102, 118

Tonsil size, 10, 11
Transform
  see Empirical transform, Empirical logistic transform, Logistic transform
Transformation
  normality, to, 119, 120
  stable effect, to, 21, 115
  stimulus, of, 110
Transition count, 73, 74

Unbiased empirical transform, 33, 34
Unbiased estimating equation, 41
Under-dispersion, 78

Variance, inflated, 77, 78
Variance, stabilization
  angular transformation, by, 28, 110

Weighted least squares, 17, 31, 40–42, 77, 78, 85–87, 106, 107, 109, 117
Weights, empirical estimation of, 31, 32, 34, 41, 42, 78
Wilcoxon test, 64